T0298779

Smart Urban Computing Applications

RIVER PUBLISHERS SERIES IN COMPUTING AND INFORMATION SCIENCE AND TECHNOLOGY

The "River Publishers Series in Computing and Information Science and Technology" covers research which ushers the 21st Century into an Internet and multimedia era. Networking suggests transportation of such multimedia contents among nodes in communication and/or computer networks, to facilitate the ultimate Internet.

Theory, technologies, protocols and standards, applications/services, practice and implementation of wired/wireless

The "River Publishers Series in Computing and Information Science and Technology" covers research which ushers the 21st Century into an Internet and multimedia era. Networking suggests transportation of such multimedia contents among nodes in communication and/or computer networks, to facilitate the ultimate Internet.

Theory, technologies, protocols and standards, applications/services, practice and implementation of wired/wireless networking are all within the scope of this series. Based on network and communication science, we further extend the scope for 21st Century life through the knowledge in machine learning, embedded systems, cognitive science, pattern recognition, quantum/biological/ molecular computation and information processing, user behaviors and interface, and applications across healthcare and society.

Books published in the series include research monographs, edited volumes, handbooks and textbooks. The books provide professionals, researchers, educators, and advanced students in the field with an invaluable insight into the latest research and developments.

Topics included in the series are as follows:-

- Artificial Intelligence
- Cognitive Science and Brian Science
- Communication/Computer Networking Technologies and Applications
- Computation and Information Processing
- Computer Architectures
- Computer Networks
- Computer Science
- Embedded Systems
- Evolutionary Computation
- Information Modelling
- Information Theory
- Machine Intelligence
- Neural Computing and Machine Learning
- Parallel and Distributed Systems
- Programming Languages
- Reconfigurable Computing
- Research Informatics
- Soft Computing Techniques
- Software Development
- Software Engineering
- Software Maintenance

For a list of other books in this series, visit www.riverpublishers.com

Smart Urban Computing Applications

Editors

M.A. Jabbar
Vardhaman College of Engineering, India

Sanju Tiwari
Universidad Autonoma de Tamaulipas, Mexico

Fernando Ortiz-Rodriguez
Tamaulipas Autonomous University Ciudad Victoria, Mexico

NEW YORK AND LONDON

Published 2023 by River Publishers
River Publishers
Alsbjergvej 10, 9260 Gistrup, Denmark
www.riverpublishers.com

Distributed exclusively by Routledge
605 Third Avenue, New York, NY 10017, USA
4 Park Square, Milton Park, Abingdon, Oxon OX14 4RN

Smart Urban Computing Applications / M.A. Jabbar, Sanju Tiwari and Fernando Ortiz-Rodriguez

Routledge is an imprint of the Taylor & Francis Group, an informa business

ISBN 978-87-7022-749-0 (print)
ISBN 978-10-0084-643-0 (online)
ISBN 978-1-003-37324-7 (ebook master)

Contents

Preface

This edited volume book is a collection of quality research articles reporting the research advances in the area of deep learning, IoT, and urban computing. This book describes new insights based on deep learning and IoT for urban computing and is useful for architects, engineers, policymakers, facility managers, academicians, and researchers who are interested in expanding their knowledge of the applications of deep learning trends involving urban computing. This book consists of 9 chapters which focus on emerging technologies and their applications to smart urban computing.

In **Chapter 1**, the authors propose a chapter on requirements analysis of data analytics software within the scope of smart university. This chapter focuses on analyzing and designing the analytical system's sustainability requirements after establishing the smart campus infrastructure. The research presents the possible outputs that can be achieved with real-time data analytics and reveals the definition, analysis, and design of the system's sustainability and development needs.

Chapter 2 presents performance analysis of deep learning models for re-identification of person in public surveillance system. This chapter proposes an automatic real-world video surveillance system model that can track and re-identify multiple persons from a single-camera tracking environment. The proposed system compares the effect of different benchmarked deep neural networks on MARS and iLIDS-VID video re-identification dataset.

Chapter 3 explores the concepts of exploiting trajectory data to improve smart city services. The goal of this chapter is to look at the way in which trajectory data mining activities are described at an existential level, the exact sort of knowledge that can be extracted through trajectory data, and the methods that trajectory data mining techniques apply toward various tasks of urban computing.

Chapter 4 discusses an end–end framework for autonomous driving cars in CARLA simulator. In this chapter, an end–end framework is proposed for autonomous driving, which processes a given image, detects traffic signals in the given image, identifies the color of the traffic lights, and then, using area of traffic signal, traffic light color, and the RGB image, predicts the steering angle, throttle, and brake values. The model can be improved by training the detection model to detect other vehicles and pedestrians and collecting more data and training on larger datasets.

In **Chapter 5**, the authors propose IoT and artificial intelligence techniques for public safety and security. This chapter discusses a number of significant technologies as well as solutions to problems that citizens face as a result of a lack of digitalization. It addresses concerns such as public infrastructure, public safety, and security, as well as providing ideal solutions. It focuses not only on AI but also on IoT, machine learning, deep learning, pattern reorganization, and big data analytics for development of a smart city that is completely functional. The analysis identifies the key barriers for widespread adoption and proposes a research path for every owing a cost-effective IoT for defense and public safety.

Chapter 6 discusses about deep learning approaches for the classification of IoT-based hyper-spectral Images. This chapter primarily deals with the application of Internet of Things and deep learning in hyper-spectral imaging analysis. For this purpose, some well-known publicly available datasets are used for the classification of different classes present in these datasets. Four main deep neural networks, especially convolution neural network, recurrent neural network (long short-term memory and gated recurrent unit), auto-encoders, and generative adversarial network have been used for the experimentation purpose. A comparative analysis of these classification techniques used for finding the accuracy is made. In the end, certain challenges in deep learning are analyzed, along with some of the emerging future research axes.

In **Chapter 7**, the authors survey on artificial intelligence and IoT for smart city. This chapter will commence with a brief introduction about artificial intelligence, AI history, benefits of AI, limitations or challenges of AI, future scope, and solutions proposed by AI. The chapter also highlights IoT, how IoT has actually made life easier and the devices to stay connected, and other advantages and challenges of AI. In this chapter, the authors discuss smart cities and shed light on various sectors like traffic management, agriculture, health care, etc., which have helped the cities get converted into smart cities.

Chapter 8 explains intelligent facility management system for self-sustainable homes in smart cities: an integrated approach. This chapter proposes a framework for an intelligent facility management system for self-sustainable homes in smart cities by integrating building information modeling (BIM), artificial intelligence (AI), Internet of Things (IoT), and big-data analytics (BDA).

In **Chapter 9**, the authors discuss emerging technology for smart living. This book chapter explores different IoT-enabled smart city applications with machine learning (ML) and deep learning (DL). Many challenges that are present in today's smart city applications are also discussed.

List of Contributors

Akkol, Ekin, *Izmir Bakircay University, Turkey*

Anand, N., *Karunya Institute of Technology and Sciences, India*

Banerjee, Anasua, *School of Computer Engineering, KIIT Deemed to be University, India*

Bhosale, Snehal A., *RMD Sinhgad School of Engineering, India*

Bijrothiya, Sadhna, *PhD Maulana Azad National Institute of Technology, India*

Demir, Yunus, *Izmir Bakircay University, Turkey*

Dogan, Onur, *Izmir Bakircay University, Turkey*

Eliiyi, Deniz Tursel, *Izmir Bakircay University, Turkey*

Garg, Bhagwati, *Union Bank of India, India*

Hiziroglu, Abdulkadir, *Izmir Bakircay University, Turkey*

Jabbar, M. A., *Professor & HoD, CSE (AI&ML), Vardhaman College of Engineering, India*

Kanaga, E. Grace Mary, *Karunya Institute of Technology and Sciences, India*

Koc, Hatice, *Izmir Bakircay University, Turkey, Gebze Technical University, Kocaeli*

Kostepen, Zeynep Nur, *Izmir Bakircay University, Turkey*

Kumar, Anil, *Government Engineering College, India*

LNC Prakash, K., *Associate Professor, CSE, CVR College of Engineering, India*

Mahor, Vinod, *IES College of Technology, India*

Nigel, K. Gerard Joe, *Karunya Institute of Technology and Sciences, India*

Pachlasiya, Kiran, *NRI Institute of Science and Technology, India*

Pandya, Vedant, *Department of Computer Engineering, Thakur College of Engineering and Technology, India*

Patil, Megharani, *Department of Computer Engineering, Thakur College of Engineering and Technology, India*

Ragha, Leena, *Ramrao Adik Institute of Technology, India*

Rawat, Romil, *Shri Vaishnav Vidyapeeth Vishwavidyalaya, Indore, India*

Saxena, Harsha, *Ramrao Adik Institute of Technology, India*

Sharma, Mukta, *Associate Professor, HOD, Department of CS & IT, TIPS, India*

Shrivastava, Shivanshu, *Department of Computer Engineering, Thakur College of Engineering and Technology, India*

Somthankar, Anuja, *Department of Computer Engineering, Thakur College of Engineering and Technology, India*

Suryanarayana, G., *Associate Professor, CSE, Vardhaman College of Engineering, India*

Swain, Satyajit, *School of Computer Engineering, KIIT Deemed to be University, India*

List of Figures

List of Tables

List of Abbreviations

AI	Artificial intelligence
AE	Auto-encoders
AV	Autonomous vehicle
AA	Average accuracy
BIRCH	Balanced iterative reduction and clustering using hierarchies
BN	Bayes network
BDA	Big-data analytics
BIF	Biologically inspired feature
BIM	Building information modeling
CARLA	Car learning to act
CCE	Categorical cross-entropy
CNN	Convolution neural network
CMC	Cumulative matching curve
DL	Deep learning
DNN	Deep neural network
DBSCAN	Density-based spatial clustering of applications with noise
DR	Dimensionality reduction
ETC	Electronic toll collection
EVA	Electronic virtual assistant
ELF	Ensemble of localized features
FDA	Fisher discriminant analysis
FLOPS	Floating-point operations per second
GRU	Gated recurrent unit
GAN	Generative adversarial networks
GIS	Geographic information systems
GPS	Global positioning systems
GSM	Global system for mobile communications
HMM	Hidden markov model
HSI	Hyperspectral imaging
iLIDS-VID	iLIDS video re-identification

IPC	Indian Penal Code
ITSD	Indoor train station dataset
IIoT	Industrial IoT
ICT	Information and communication technology
IP	Ingress protection
IoT	Internet of things
IHTLS	Iterative hankel total least squares
KA	Kappa accuracy
KISSME	Keep-it-simple-and-straightforward metric
KPI	Key performance indicator
K-NN	K-nearest neighbor
KDD	Knowledge discovery in databases
LiDAR	Light detection and ranging
LDFV	Local descriptors encoded by Fisher vectors
LOMO	Local maximal occurrence
LR	Logistic regression
LSTM	Long short-term memory
ML	Machine learning
mAP	Mean average precision
MSE	Mean squared error
MARS	Motion analysis and re-identification set
NB	Naive bayesian
NCRB	National crime records bureau
NHS	National hurricane service
NLP	Natural language processing
OPTICS	Ordering points to identify the clustering structure
OA	Overall accuracy
PCCA	Pairwise constrained component analysis
PCA	Principal component analysis
RFID	Radio-frequency identification
RNN	Recurrent neural network
RGB	Red, green, and blue
ReID	Re-identification
REMO	Relative motion
RMSE	Root mean squared error
SDLC	Software development lifecycle
SLL	Special and local laws

SAE Stacked auto-encoder network
SVM Support vector machine
UAV Unmanned aerial vehicle
WSN Wireless sensor network
YOLO You only look once

1

Requirement Analysis of Data Analytics Software Within the Scope of a Smart University

Ekin Akkol[1], Hatice Koc[1,2], Onur Dogan[1], Zeynep Nur Kostepen[1], Yunus Demir[1], Abdulkadir Hiziroglu[1], and Deniz Tursel Eliiyi[1]

[1]Izmir Bakircay University, Turkey
[2]Gebze Technical University, Kocaeli
Email: onur.dogan@bakircay.edu.tr

Abstract

As small cities, the primary purpose of smart universities is to facilitate campus life by applying technology and use scarce resources more effectively by reducing consumption. Unlike other smart university studies, this research focuses on analyzing and designing the sustainability requirements of an analytical system after establishing the smart campus infrastructure. The research presents the possible outputs that can be achieved with real-time data analytics and reveals the definition, analysis, and design of the sustainability and development needs of the system. The system outputs are produced in three main groups such as descriptive, predictive, and prescriptive. They are presented to the service of relevant stakeholders in universities in an easily accessible way. The types of data required to obtain these outputs are also investigated in detail. The V-model system development methodology was followed in the project.

1.1 Introduction

Internet and communication technologies (ICT) help significantly in improving living conditions, using scarce resources productively, providing energy efficiency, enhancing management services [1], and protecting the environment in urban areas. These technologies facilitate extracting information from real-time data and controlling remote devices with sensors and end systems in a transparent and seamless manner [2]. A smart city is built by integrating these technologies in order to manage hospitals, power grids, railways, bridges, tunnels, roads, buildings, water systems, dams, oil and gas pipelines, and other systems. The concept was evolved to tackle problems stemming from daily livelihoods, environmental protection, public safety, and city services, as well as industrial and commercial activities [3].

The concept of a smart city has also impacted the emergence of the concept of smart universities since a smart university can be acknowledged as a smart city [1]. A smart university is a system that is improved as green, robust, personalized, responsible, interactive, adaptive, and accessible and allows knowledge to be accessed seamlessly [4]. The concept of a smart university is based on enhancing the quality of educational, research, commercial, and other university processes by utilizing ICT, smart technologies, smart features, smart software and hardware systems, smart devices, smart pedagogy, smart curricula, smart learning, and academic analytics [5]. Radio frequency identification (RFID) and wireless sensor network (WSN) technologies are employed in controlling processes and increasing energy efficiency [4].

The smart university enables to extract information and knowledge from raw data automatically. Self-description, self-discovery, self-optimization, and self-organization are features of a smart university. All of these features help change business functions and existing knowledge, experience, or behavior in order to enhance operations and performance. These features also help in making a prediction about events to increase the quality of education, service management, and energy efficiency and using limited resources productively. The smart campus aims to establish a framework for digitalizing processes. For example, on smart campus, all campus-related information and courses can be accessed remotely and locally [6].

Smart universities should improve software systems for facilitating lower energy consumption and waiting times in campus facilities, enhancing personnel productivity and education quality, participating in the process of problem identification and resolution at the university, and analyzing the usage of resources [7]. Thus, it has been planned to improve a real-time data analytics system for ensuring the sustainability of a smart university in this study. This system will use descriptive, predictive, and prescriptive analytics to provide reports for stakeholders in the university. The V-model system development methodology was employed to design the system. The V-model system development is a type of waterfall methodology that requires verification and validation of the system at every stage [1]. It needs to perform requirement analysis as the software project is to be completed after grasping the requirements completely and documenting all requirements in detail [8].

A software is formed using computer programs, associated documentation, and configuration data [9]. Therefore, exploring and meeting user needs correctly when developing a software are significant to take advantage of time, budget, and the quality of the system [8]. Requirement analysis has a substantial influence on executing a software development lifecycle. It helps explore and solve missing, ambiguities and disagreements about a software system [10]. It is performed to determine the functions, constraints, and goals of the system in detail, which gives information about what the system should do, which constraints will occur when operating and implementing the software system, and which functions and services from the system will be required by the user [9]. It helps design and schedule projects and allocate resources. It is performed with several phases which are composed of determining, examining, confirming, prioritizing, and documenting requirements as well as identifying requirements changes [8]. The requirements can be split into two categories which are functional requirements and non-functional requirements. Functional requirements reveal how the reactions of the software system should be to input and the behaviors of the software system for specific situations [9]. That is, they describe the behaviors and actions of the system users [8]. However, the constraints of the software system are depicted with non-functional requirements [9], which can arise from the platform, environment, design limitations, or dependencies of the system [8].

The university has created a methodological framework that is formed of three stages consisting of a single facility, extended facility and environment, and smart campus, which is displayed in Figure 1.1. In Stage 1, the focus is on energy, environment, and classroom domains while it is extended with training domain, administrative feedback domain, environment domain, and availability domain in Stage 2. However, a data-oriented approach is embraced, and it is planned to utilize predictive, prescriptive, and descriptive analytics in Stage 3. Thus, Stage 3 focuses on the data obtained from the systems in Stage 1 and Stage 2 and external sources, and contains real-time analytics, monitoring, reporting, and performance measurement domain. A data-collection-recording server and web-based software are utilized to store and extract information and knowledge and make predictions. The information and knowledge help develop a user-friendly and dynamic reporting system that enables to create a sustainable campus and thrive physical conditions to control and manage university [7].

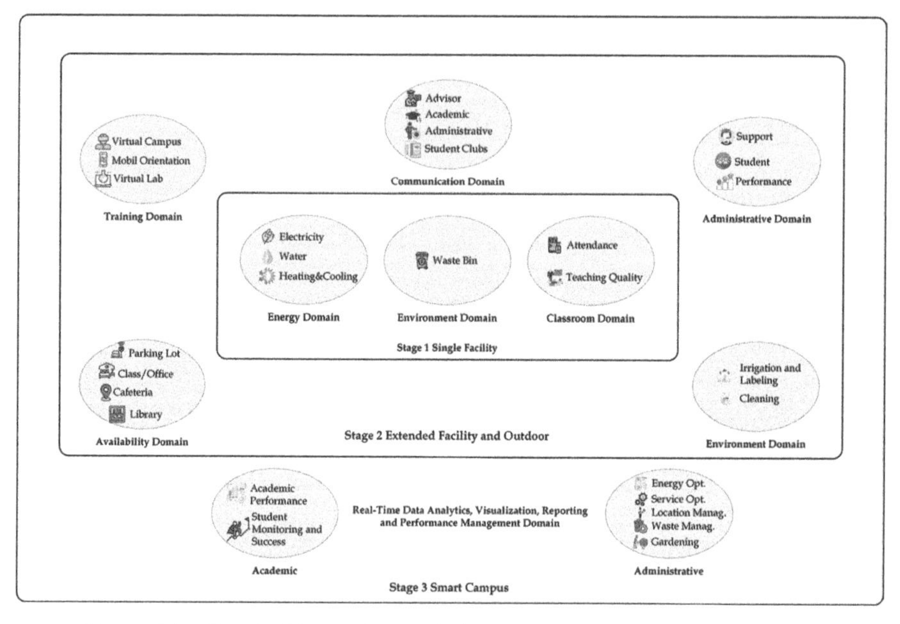

Figure 1.1 Sustainable and data analytics oriented smart campus framework.

This book chapter was prepared to describe, analyze, and design the sustainability requirements of the analytical system after establishing a smart campus infrastructure and present possible outputs that can be achieved with real-time data analytics. The rest of this chapter is organized as four sections. In Section 1.2, smart campus studies and requirements analysis studies are discussed and summarized. The utilized methodology is explained in Section 1.3. A flowchart for software requirements analysis for real-time data analytics within the scope of smart university is also shown in this section. In Section 1.4, several examples are provided, consisting of domains, sub-domains, technology, and benefits of real-time data analytics framework for smart university. Besides, the requirement analysis for this real-time data analytics system is exemplified in this section. The chapter is finalized with a conclusion in Section 1.6.

1.2 Literature Review

Research and projects in the literature on smart university applications focus on the application of high-level smart abilities with different studies in terms of applying the concept of smart university to learning activities. In order to examine these studies in detail, a literature review was performed. Since this study deals with the requirements analysis of data analytics software within the scope of smart university, the literature review has been made under two separate subsections, namely smart university and requirements analysis.

First, smart campus studies in the literature were examined. These studies are classified in the subfields of education, energy, and administrative, where smart campus studies are mostly carried out and given in this way. In the second part of the literature review, the usage areas of the requirements analysis to be used within the scope of the study are given in detail. The literature in this area is also classified into software, business, and system subfields, where requirements analysis is used more frequently.

1.2.1 Smart campus studies

Studies and projects on smart university applications offer different perspectives on the concept of smart university and its applications. When the studies in the literature are examined, it can be observed that the majority focused on high-level applications by focusing on different fields such as education, energy, and administrative of universities [11]. For example, in a study conducted in the field of smart university, the authors wanted to include the concept of smart in universities and suggested various applications in classroom support, security, and education [12]. In some studies, the concept of smart city was discussed, and smart systems were applied to universities in terms of being a small city. In this area, IoT technologies have been used to enable remote access to universities and increase traceability [13].

Most studies aimed at increasing efficiency in education, energy, and administrative fields. When these areas are investigated, the majority are in the field of education. In order to increase the quality of education, smart systems that can be employed in the classroom have been proposed [14]. For example, a study integrated smart systems into classrooms to control devices such as lights, smart boards, projectors, air conditioners, doors, and temperature sensors in classrooms, thus making it possible to control all devices used in the classroom from a single place. In addition, this system also provided remote access to the classrooms and showed that energy-consuming devices such as air conditioners and projectors, which remain on when the classrooms are not in use, can be detected and energy savings can be achieved [15].

Integrating the smart concept into campuses also makes distance learning possible. In [12], several opportunities for innovations in education are provided by integrating the concept of smart university into classrooms.

Moreover, game-based approaches [16], multimedia conferences [17], and the effectiveness of smart phone applications and smart campus applications in education [18] were investigated.

The energy field is another field of great importance for universities. Efforts to save energy in order to ensure the effective use of limited resources provide great advantages. For example, one study developed a web-based energy management system that can be integrated into various usage areas within the campus in their study at the Technical University of Crete in the field of energy [19]. In this way, they have enhanced energy efficiency by achieving a 30% energy savings throughout the campus. In another study in the field of energy, it was possible to remotely control the equipment utilized in the classrooms on the campus, and thus to turn off electrical devices such as projection, smart board, light, and air conditioner when not in use [15]. Thus, significant energy savings were acquired. In another study [20], the authors presented the importance of improvements made in the field of electricity in campus buildings with energy analysis.

Universities are units where control and communication are crucial. Both the planning of targeted strategies and the good interaction and communication between units ensure that the processes are carried out more accurately and clearly. For this reason, the fact that smart systems can be applied in the field of management within the university will improve the control system as well as strengthen communication. For example, a study in the administrative field suggested a new approach for an IoT-based employee performance appraisal to avoid manual performance appraisal of employees. Thanks to this system [21], the study obtained results about the performance of the employees and argued that there could be an increase in performance in this way. Another study used IoT technologies to enable educators and administrators to turn data into action in the field of education [22]. With the automatic system created, student performance evaluation has become easier. In [23], the authors aimed to present the results to users more effectively as well as the communication of objects with each other via the use of IoT technologies. For this reason, an interaction has been established between smart objects and users.

Table 1.1 shows the studies in the field of smart universities in the literature and the main areas covered in these studies.

Various methods have been used in smart university studies. In addition to these methods, a requirements analysis at the beginning of the work is critical in analyzing the benefits offered at the end of the studies and choosing the technologies that should be employed. With a requirements analysis, it becomes possible to use limited resources more effectively. It is also possible to provide the needs of the units within the university and the necessary data sources with this method.

Table 1.1 Previous studies and main areas of study.

Areas	Studies
Education	[12, 14–18, 24–30]
Energy	[15, 19, 20, 28, 29, 31–34]
Administrative	[21–24, 30]

1.2.2 Requirements analysis studies

Requirements analysis is a concept that many software requirements arise within the concept of smart university. Using smart systems in campus areas and integrating software into processes enables control and accessibility. For this reason, it is significant to determine the software correctly. The correct software selection and the effective application of the requirements analysis provide a suitable roadmap for achieving goals.

Software selection consists not only of the programs but also of the design and philosophies expected of the programs. Therefore, software requirements analysis is crucial in meeting the expectations of the selected software. For software systems to be acceptable, the requirements must be well defined, analyzed, and prioritized. A well-executed requirements analysis also ensures the effective use of limited resources and budget [35]. For example, one study utilized requirements analysis to identify the business strategy and IT requirements of an organization [36]. Another one used requirements analysis to reveal software requirements [37]. The authors also presented the most recent or prominent techniques to tackle the problems. In [38], information requirements analysis was utilized for the development of analytical information systems. Thus, the research was conducted for uncovering, verifying, and managing information requirements.

Requirements analysis has a great impact on determining the requirements, making the plans correctly, and controlling the budget, time, and resources [39]. Besides, the probability of rejection of the software selected through a requirements analysis is lower. The reason for this is that all options are evaluated with this analysis and the right decisions are taken. When the literature is examined, it is seen that many studies have been carried out in the field of requirements analysis. The most common requirements analysis fields in the literature were observed as software, business, and system. For this reason, in this study, a classification is made between these three groups. Among the studies examined, it was seen that the most frequently used technique was software requirements analysis. These are various studies on IT technologies, big data, cloud computing, and data analysis; and all of these areas are combined under software requirements analysis. Table 1.2 shows the classification of the studies reviewed in the literature.

Table 1.2 Classification of requirements analysis studies reviewed in the literature.

Areas	Studies
Software	[37, 40–53]
Business	[36, 38, 54, 55]
System	[56–58]

In the literature review on requirements analysis, it has been seen that this analysis is frequently employed in the software field. The reason for this is that determining the software requirements correctly draws the right path to the studies. As information technologies advance, various software are used in all sectors and fields. The use of IoT technologies and, accordingly, information technologies in the smart university concept will bring along powerful software. In the studies carried out in the field of smart university until now, usually a new application idea has been put forward rather than developing a model. Within the scope of this study, a software requirements analysis was applied for the smart university.

1.3 Proposed Methodology

1.3.1 Software development process

Software development activities should be performed by stakeholders in order to produce a software product. Since it is not an ideal method, each team has developed the most suitable process for their own structure. Despite these differences, there are some common activities in all software development processes:

- determining the features of the software;
- identifying software needs and limitations;
- realization of the software;
- control of the software after delivery to the customer;
- revision of the software according to the feedback from the field [59].

1.3.2 Software development models and its steps

Software development lifecycle models can be employed for software development processes. Some of them are as follows.

No planning should be performed for an *ad hoc* model. Even if planning is done, transactions tend to take place unplanned.

Waterfall model can be considered as the ancestor of software development processes. It includes the phases of requirements analysis, design, coding, integration, operation, and maintenance.

Incremental model is based on progress by dividing the system into modules rather than designing it all at once. In this model, feedback is continuously received from the user. This is the factor that increases satisfaction.

The main purpose of the spiral model is to reduce risks. Each stage of the model is designed to decrease risks [60, 61].

The software development lifecycle (SDLC) is a process that can return to where it started. The stages of this cycle are as follows [61]:

- Planning
- Identification
- Designing
- Development
- Integration and tests
- Implementation
- Maintenance and evaluation

1.4 V-Model

The steps of this model are planning, requirements analysis, design, coding, tests, and maintenance. It is called V-model because the steps form the letter V. In the V-model, there are steps that are on the same level. In the model, first, planning, requirements analysis, design, and coding are performed. Once the coding step is completed, each test is performed. After design is performed in detail, unit tests for each module are done. Then, integration testing process checks the compatibility of these modules with each other. The requirements analysis is conducted with users, and then whether the requirements analysis is suitable is tested. Finally, there are planning and maintenance. All steps were not applied in this study. The software requirements step in the V-model is performed [61].

1.4.1 Previous steps of the project

Smart campus project is conducted as three different phases which are "A V-Model Software Development Application for Sustainable and Smart Campus Analytics Domain" [62], "Sustainable and Data Analytics Focused

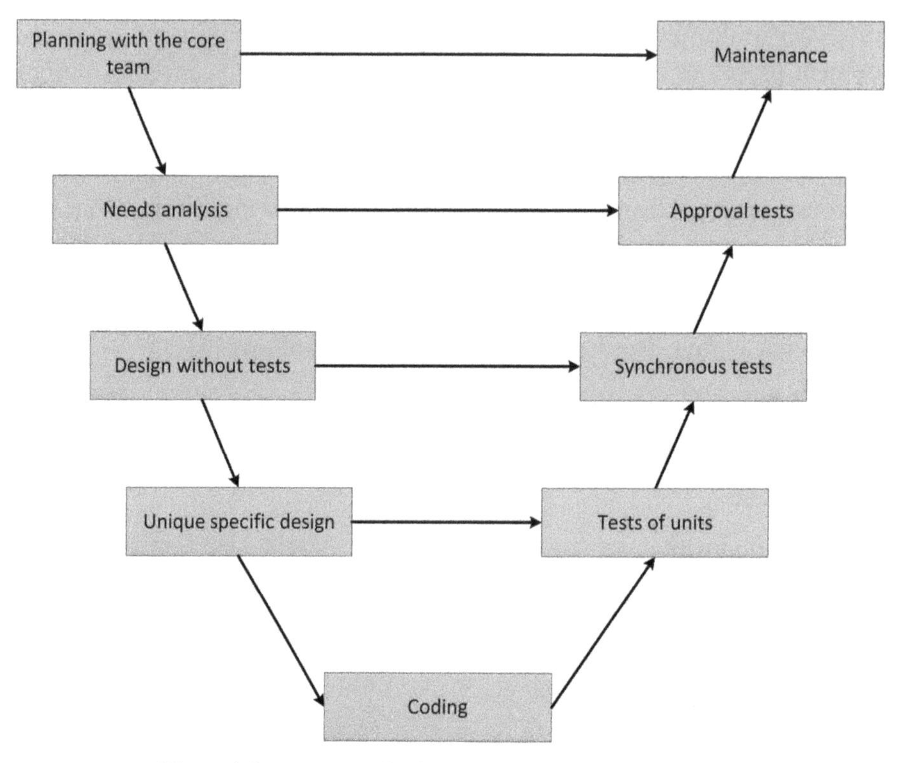

Figure 1.2 V-model of software requirements specification.

Campus Project Feasibility Report" [63], and "The Software Requirements Specifications." These phases have not been consecutive.

1.4.2 V-model of software requirements specifications

Planning and design are performed until coding. After coding, the steps of testing synchronization and maintenance are applied. The steps testing synchronization and maintenance are carried out during the steps of planning and design.

1.4.3 Planning with the core team

This step is the beginning of all the steps of the project. Parameters such as the feasibility and benefits of the project are revealed. For these reasons, it is the most crucial step of the project [61].The main subjects and areas within the scope of the project were planned with the core team and an intellectual infrastructure was established [62].

The main framework of the software was determined with the core team. Then, interviews were conducted with relevant interlocutors to identify the key performance indicators (KPIs) of all stakeholders who will use the system. The indicators were presented to the stakeholders and their feedback was obtained.

1.5 Adaptation of the Proposed Methodology

In the process of determining the software requirements analysis of the smart university project, it was first revealed which units are responsible for the main domains and sub-domains related to the different phases. In addition, it has been determined what kind of benefits will be obtained in the studies of the fields and which technologies will be used. In this way, data sources and relevant units of the university were determined. Table 1.3 presents examples from the report created on the main domains, sub-domains, stage information, benefits, equipment to be used, and related units.

There are many different data sources in the framework established within the scope of the smart university project. The data coming from these data sources will be backed up in databases and displayed in real time to the relevant people and units on the dashboard with descriptive, predictive, and prescriptive reports. The framework prepared in this context is shown in Figure 1.3.

It is crucial that the reports prepared for the relevant persons and the units are usable and useful. Analyzing and reporting the data obtained from various data sources by processing according to the needs will positively affect the decision-making processes. The study was carried out step by step with the establishment of the model. Many different factors were taken into consideration while performing the requirements analysis. First, the study team identified possible needs for all departments. During the determination of needs, the type of the report was first discussed. Descriptive, predictive, and prescriptive types of reports were included within the scope of requirements analysis. Along with the types of analytics report, the expected benefit from the report, the requirement scores for the report, the report production frequency, the report name, the description of the report, and the data that should be included in the report were included. A draft was created by the work team by creating sample reports for each department. It has also been discussed which technologies will be used to determine software requirements.

Next, the draft studies were presented to the senior managers in all departments of the university and the needs were tried to be determined as a result of the meetings. The most critical point here is to determine the reports

Table 1.3 Matching domains and departments.

Domain	Sub-Domain	Stage	Benefits	Technologies	Related Departments
Energy domain	Electricity	1	Reducing electricity consumption with sensors	Presence sensor and photocell	Department of Construction and Technical Works
Energy domain	Heating and cooling	1	Reducing air conditioning costs with sensors to be placed in closed areas	Temperature sensor and humidity sensor	Department of Construction and Technical Works
Environment domain	Bin	1	To optimize the work of cleaning personnel by determining the fullness of the garbage cans	Occupancy sensors for large garbage areas, kiosk, or mobile applications for small garbage	Administrative Financial Affairs Department
Administrative domain	Staff–student entry and exit tracking	1	Taking attendance with card reading devices at the entrance and exit doors	Card readers and turnstiles	Student Affairs Department, Personnel Department, Administrative Financial Affairs Department, and Academic Units
Contact domain	Academic	2	Strengthening the communication channel between students and academic staff	Web and mobile applications	Academic Units, Student Representatives, and Student Affairs Department
Administrative domain	Energy optimization	3	Heating/cooling and lighting reports, energy consumption forecast for the coming periods, and achieving a goal under certain constraints to find the optimum uptime	Web and mobile applications	Department of Construction and Technical Works, Administrative Financial Affairs Department, and Academic Units

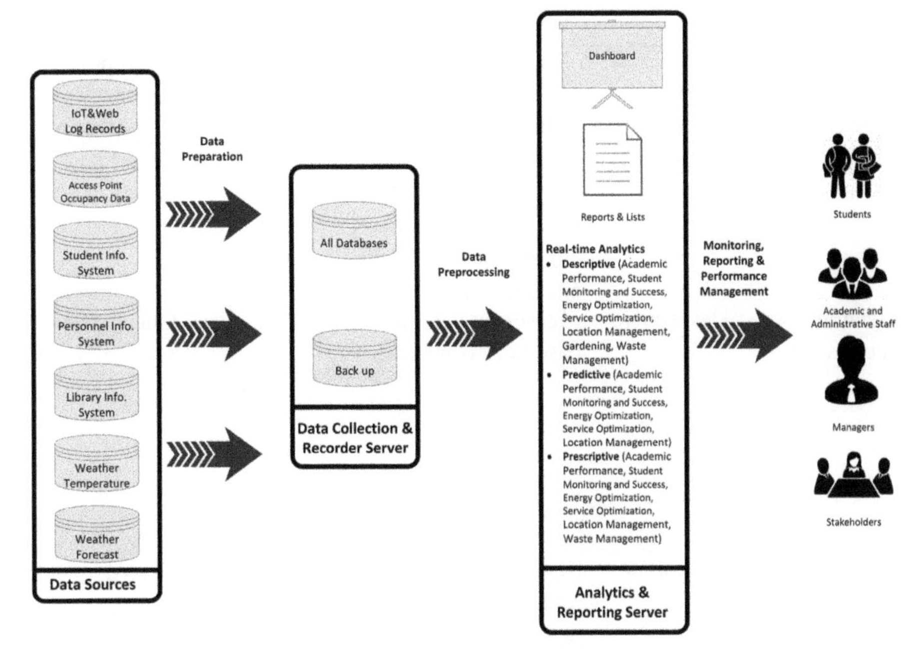

Figure 1.3 Real-time analytics framework [1].

that are suitable for the needs, sustainable, and useful. It has been tried to include as much detail as possible in the reports created during the determination of the needs. Specific topics such as chart types in the reports on the dashboard were also discussed. A few examples of requirements analysis are shown in Table 1.4.

In the reports shown in Table 1.4, reports serving different purposes were created from almost every field of the university. All reports created were approved by experts and were stated as beneficial.

While creating Table 1.4, many issues such as the benefits of the reports, the data to be included in the reports, and the necessity of the reports were studied in detail. In addition, information was provided as to how the reports would be displayed on the dashboard. With the feasibility study and the approval of end users and experts, the software needs have been fully determined for the smart university.

1.6 Conclusion

Many studies of smart university aim to raise productivity in the field of energy, education, and administration because smart university has positive

Table 1.4 Requirement analysis reports.

Related unit	Analytics type	Expected benefit	Requirement score (0–10)	Report generation frequency
Department of library	Prescriptive	Making recommendations taking into account the current habits of staff and students.	8	25 per day
General secretariat	Descriptive	Measuring person-based performance.	7	Once a week
Department of health, culture and sports	Predictive	Reducing waiting times by directing people to the empty dining hall.	6	Once a month
Rectorate & general secretariat	Descriptive	Graphical monitoring of student numbers on the basis of faculty/ department by the senior management.	9	Once a day
Academic units	Descriptive	To increase the success of students, lecturers, and university.	8	Four per year
Department of construction and technical works	Prescriptive	Fast and effective action will be taken by identifying the area with unusual consumption.	7	Six per day

Report name	Report description	Data required in the report
Recommendation system	Creating recommendation systems	Book code, type of book, number of borrowings, name of borrowers, department, and faculty
Administrative staff performance	Seeing the jobs assigned to the personnel and measuring their performance. KPIs should be determined by the relevant units. A report will be created containing the personal and manager's evaluation	Staff name, unit, working time at the university, daily average time at university, and data in KPIs
Dining hall use expectation	Estimating the density of the next month according to the current density in the dining halls	Dining hall number, dining hall name, dining hall capacity, date, number of people, number of personnel, and number of students
Student distribution report	Number of students based on department/faculty	Name of faculty, number of male, female, Turkish citizens, number of foreigners, number of active students, number of graduates, and number of resigned students (department name and city distribution)
Student attendance reports	Determination of student reports (course attendance reports, course success graph, and student success/failure) according to courses	Student name, department, class, course attendance report, and course success graph
Unusual electricity consumption tracking	Determining and reporting the area where action should be taken according to the electricity consumption data of 30 different points and consumption anomalies	Electricity consumption data of 30 different points [electricity consumption of commercial facilities (base stations, cafes, stationery, etc.), main consumption point (whole facility), tracking of corridors, system room, lab, etc.]

effects on the quality of education, energy-consuming, and control and communication with/within units, personnel, and students in a university. Izmir Bakircay University has implemented a smart university project encompassing all these fields. In this book chapter, the requirements analysis of a real-time data analytics system for smart university was explained and evaluated. This real-time data analytics system presents descriptive, predictive, and prescriptive reports for all departments in Izmir Bakircay University. It is the last phase in the sustainable and smart campus framework improved by Izmir Bakircay University and helps make decision and prediction in all departments of the university.

A V-model software development methodology was employed when the real-time data analytics system was developed. The V-model software is a type of waterfall methodology, which gives importance to verification and validation of the software and is employed if i) all features are ready before starting the project, ii) the systems need highly reliable software, and iii) technology and alternative solutions are identified.

A requirements analysis was needed by this methodology for understanding the requirements from early stage to completing the software project and documenting all requirements in detail. Besides, it was crucial to determine the requirements for software at the beginning in order to control resources, budget, and time and advance the quality of a system. A well-organized requirements analysis provided a proper roadmap to adopt smart systems and integrate software into processes in campus and influence SDLC positively.

When identifying software requirements analysis for the real-time data analytics system in the project, some features such as relevant units, main domains, sub-domains, the types of analytics report, the expected benefits, the description of the report, and the data included in the report were analyzed for each department in the university with the senior managers of these departments.

Personnel in all departments and project team believe that the reports provided by this real-time data analytics system assist to use limited resources effectively, save time and energy, and improve performance while controlling processes and performance in the university. Therefore, in the following stages of the project, the requirements analysis will be utilized and evolved when adapting real-time data analytics system according to the needs and reviews of the stakeholders.

1.7 Acknowledgment

This work has been supported by Izmir Bakırcay University Scientific Research Projects Coordination Unit, under grant number GDM.2021.006.

References

[1] O. Dogan, S. Bitim and A. Hiziroglu, "A V-Model Software Development Application for Sustainable and Smart Campus Analytics Domain," *Sakarya University Journal of Computer and Information Sciences,* vol. 4, no. 1, pp. 111–119, 2021.

[2] T.-H. Kim, C. Ramosand S. Mohammed, "Smart City and IoT"*Future Generation Computer Systems,* vol. 76, no. 2, pp. 159–162, 2017.

[3] K. Su, J. LiandH. Fu, "Smart city and the applications"*2011 International Conference on Electronics, Communications and Control (ICECC),* Ningbo, 2011.

[4] C. Heinemann and V. L. Uskov, "Smart university: literature review and creative analysis"*Smart universities: concepts, systems, and technologies*, Cham, Springer, 2017, pp. 11–46.

[5] V. L. Uskov, J. P. Bakken, R. J. Howlett and L. Jain, "Innovation in smart universities"*Smart universities: concepts, systems, and technologies*, Cham, Springer, 2017, pp. 1–7.

[6] V. L. Uslov, J. P. Bakken, K. Srinivas, A. V. Uskov, C. Heinemann and R. Rachakonda, "Smart university: conceptual modelling and systems' design"*Smart universities: concepts, systems and technologies*, Cham, Springer, 2017, pp. 49–86.

[7] Z. N. Kostepen, E. Akkol, O. Dogan and A. Hiziroglu, "A Framework for Sustainable and Data-driven Smart Campus"*International Conference on Enterprise Information Systems (ICEIS 2020)*, 2020.

[8] S. T. Demirel and R. Das, "Software requirement analysis: Research challenges and technical approaches"*2018 6th International Symposium on Digital Forensic and Security (ISDFS)*, Antalya, 2018.

[9] I. Sommerville, "Requirements engineering"*Software engineering*, Boston, Pearson, 2011, pp. 82-117.

[10] A. A. Lopez-Lorca, G. Beydoun, R. Valencia-Garcia and R. Martinez-Bejar, "Supporting agent oriented requirement analysis with ontologies"*International Journal of Human-Computer Studies,* cilt 87, pp. 20–37, 2016.

[11] M. Alvarez-Campana, G. Lopez, E. Vazquez, V. A. Villagra and J. Berrocal, "Smart CEI Moncloa: An IoT-based Platform for People Flow and Environmental Monitoring on a Smart University Campus," *Sensors,* p. 2856, 2017.

[12] A. Majeed and M. Ali, "How Internet-of-Things (IoT) making the university campuses smart? QA higher education (QAHE) perspective," in *8th Annual Computing and Communication Workshop and Conference*, 2018.

[13] P. Sotres, J. R. Santana, L. Sánchez, J. Lanza ve L. Muñoz, «Practical Lessons From the Deployment and Management of a Smart City Internet-of-Things Infrastructure: The SmartSantander Testbed Case, » *IEEE Access,* no. 5, pp. 14309–14322, 2017.
[14] P. Boulanger, "Application of augmented reality to industrial teletraining," in *First Canadian Conference on Computer and Robot Vision*, 2004.
[15] L.-S. Huang, J.-Y. Su and T.-L. Pao, "A Context Aware Smart Classroom Architecture for Smart Campuses," *Applied Sciences,* p. 1837, 2019.
[16] X. Zhai, Y. Dong and J. Yuan, "Investigating Learners' Technology Engagement - A Perspective From Ubiquitous Game-Based Learning in Smart Campus," *IEEE Access,* pp. 10279–10287, 2018.
[17] W. Zhang, X. Zhang and H. Shi, "MMCSACC: A Multi-Source Multimedia Conference System Assisted by Cloud Computing for Smart Campus," *IEEE Access,* pp. 35879–35889, 2018.
[18] Y.-B. Lin, L.-K. Chen, M.-Z. Shieh, Y.-W. Lin and T.-H. Yen, "CampusTalk: IoT Devices and Their Interesting Features on Campus Applications," in *IEEE*, 2018, pp. 26036–26046.
[19] D. Kolokotsa, K. Gobakis, S. Papantoniou, C. Georgatou, N. Kampelis, K. Kalaitzakis, K. Vasilakopoulou and M. Santamouris, "Development of a web based energy management system for University Campuses: The CAMP-IT platform," in *Energy and Buildings*, 2016, pp. 119–135.
[20] C. J. Kibert, Sustainable construction: green building design and delivery, John Wiley & Sons, 2016.
[21] N. Kaur and S. K. Sood, "A game theoretic approach for an IoT-based automated employee performance evaluation," *IEEE Systems Journal,* pp. 1385–1394, 2015.
[22] P. Verma, S. K. Sood and S. Kalra, "Smart computing based student performance evaluation framework for engineering education," *Computer Applications in Engineering Education,* pp. 977–991, 2017.
[23] M. Alessi, E. Giangreco, M. Pinnella, S. Pino, D. Storelli, L. Mainetti, V. Mighali and L. Patrono, "A Web based Virtual Environment as a connection platform between people and IoT," in *International Multidisciplinary Conference on Computer and Energy Science*, 2016.
[24] N. Gligoric, T. Dimcic, S. Krco, V. Dimcic, J. Vaskovic and I. Vojinovic, "Internet of Things Enabled LED Lamp Controlled by Satisfaction of Students in a Classroom," *A publication of IPSI Bgd Internet Research Society New York,* 2014.
[25] M. Leisenberg and M. Stepponat, "Internet of Things Remote Labs: Experiences with Data Analysis Experiments for Students Education," in *IEEE Global Engineering Education Conference*, 2019.

[26] G. Vignali, M. Bertolini, E. Bottani, L. D. Donata, A. Ferraro and F. L. Longo, "Design and testing of an augmented reality solution to enhance operator safety in the food industry," *International Journal of Food Engineering,* 2017.

[27] P. Horejsi, "Augmented reality system for virtual training of parts assembly," *Procedia Engineering,* pp. 699–706, 2015.

[28] Y.-W. Wu, L.-M. Young and M.-H. Wen, "Developing an iBeacon-based ubiquitous learning environment in smart green building courses," *The International journal of engineering education,* pp. 782–789, 2016.

[29] W.-J. Shyr, L.-W. Zeng, C.-K. Lin, C.-M. Lin and W.-Y. Hsieh, "Application of an energy management system via the internet of things on a university campus," in *Eurasia Journal of Mathematics*, 2018.

[30] J. C. Cha and S. K. Kang, "The study of a course design of iot manpower training based on the hopping education system and the esic program," *International Journal of Software Engineering and Its Applications,* pp. 71–82, 2015.

[31] S. Habibi, "Smart innovation systems for indoor environmental quality (IEQ)," *Journal of Building Engineering,* pp. 1–13, 2016.

[32] M. Jain, N. Kaushik and K. Jayavel, "Building automation and energy control using IoT - Smart campus," in *2nd International Conference on Computing and Communications Technologies*, 2017.

[33] M. A. Al Mamun, M. A. Hannan, A. Hussain and H. Basri, "Wireless sensor network prototype for solid waste bin monitoring with energy efficient sensing algorithm," in *IEEE 16th International Conference on Computational Science and Engineering*, 2013.

[34] K. Akkaya, I. Guvenc, R. Aygun, N. Pala and A. Kadri, "IoT-based occupancy monitoring techniques for energy-efficient smart buildings," in *IEEE Wireless communications and networking conference workshops*, 2015.

[35] P. Achimugu, A. Selamat, R. Ibrahim and M. N. Mahrin, "A systematic literature review of software requirements prioritization research," *Information and software technology,* vol. 56, no. 6, pp. 568–585, 2014.

[36] S. J. Bleistein, K. Cox and J. Verner, "Strategic alignment in requirements analysis for organizational IT: an integrated approach," in *In Proceedings of the 2005 ACM symposium on Applied computing*, 2005.

[37] B. Davey and K. R. Parker, "Requirements elicitation problems: a literature analysis," *Issues in Informing Science and Information Technology,* vol. 12, pp. 71–82, 2015.

[38] F. Stroh, R. Winter and F. Wortmann, "Method support of information requirements analysis for analytical information systems," *Business & Information Systems Engineering,* vol. 3, no. 1, pp. 33–43, 2011.

[39] D. Leffingwell ve D. Widrig, Managing software requirements: a unified approach, Addison-Wesley Professional, 2000.

[40] A. I. Anton, "Goal-based requirements analysis," in *In Proceedings of the second international conference on requirements engineering*, 1996.

[41] A. Al-Ali, I. A. Zualkernan, M. Rashid, R. Gupta and M. AliKarar, "A smart home energy management system using IoT and big data analytics approach," in *Transactions on Consumer Electronics*, 2017.

[42] H. H. Altarturi, K.-Y. Ng, M. I. H. Ninggal, A. S. A. Nazri and A. Azim, "A requirement engineering model for big data software," in *Conference on Big Data and Analytics (ICBDA)*, 2017.

[43] J. Al-Jaroodi and N. Mohamed, "Characteristics and requirements of big data analytics applications," in *2nd International Conference on Collaboration and Internet Computing (CIC)*, 2016.

[44] N. H. Madhavji, A. Miranskyy and K. Kontogiannis, "Big picture of big data software engineering: with example research challenges," in *IEEE/ACM 1st International Workshop on Big Data Software Engineering*, 2015.

[45] Z. Radovilsky, V. Hegde, A. Acharya and U. Uma, "Skills requirements of business data analytics and data science jobs: A comparative analysis," *Journal of Supply Chain and Operations Management,* vol. 16, no. 1, pp. 82–101, 2018.

[46] R. R. Lutz, "Targeting safety-related errors during software requirements analysis," *Journal of Systems and Software,* vol. 34, no. 3, pp. 223–230, 1996.

[47] L. Fernandez-Sanz and S. Misra, "Analysis of cultural and gender influences on teamwork performance for software requirements analysis in multinational environments," *IET software,* vol. 6, no. 3, pp. 167–175, 2012.

[48] K. Verma and A. Kass, "Requirements analysis tool: A tool for automatically analyzing software requirements documents," in *In International semantic web conference*, Berlin, Heidelberg, 2008.

[49] J. Bendík, "Consistency checking in requirements analysis," in *In Proceedings of the 26th ACM SIGSOFT international symposium on software testing and analysis*, 2017.

[50] C. Hösel, C. Roschke, R. Thomanek, T. Rolletschke, B. Platte and M. Ritter, "Process Automation in the Translation of Standard Language Texts into Easy-to-Read Texts–A Software Requirements Analysis," in *In International Conference on Human-Computer Interaction*, 2020.

[51] S. Zillner, N. Lasierra, W. Faix and S. Neururer, "User needs and requirements analysis for big data healthcare applications," *In e-Health–For Continuity of Care,* pp. 657–661, 2014.

[52] P. Gölzer, P. Cato and M. Amberg, "Data processing requirements of industry 4.0-use cases for big data applications," in *ECIS 2015 Research-in-Progress Papers*, 2015.

[53] A. Ismail, H.-L. Truong and W. Kastner, "Manufacturing process data analysis pipelines: a requirements analysis and survey," *Journal of Big Data,* vol. 6, no. 1, pp. 1–26, 2019.

[54] J. L. d. l. Vara, J. Sánchez and Ó. Pastor, "Business process modelling and purpose analysis for requirements analysis of information systems," in *In International Conference on Advanced Information Systems Engineering*, Berlin, Heidelberg, 2008.

[55] A. G. Sutcliffe, "Requirements analysis for socio-technical system design," *Information Systems,* vol. 25, no. 3, pp. 213–233, 2000.

[56] R. J. Boland, "Control, causality and information system requirements," *Accounting, Organizations and Society,* vol. 4, no. 4, pp. 259–272, 1976.

[57] C. Potts, K. Takahashi and A. I. Anton, "Inquiry-based requirements analysis," *IEEE software,* vol. 11, no. 2, pp. 21–32, 1994.

[58] R. Vidgen, "Stakeholders, soft systems and technology: separation and mediation in the analysis of information system requirements," *Information Systems Journal,* vol. 7, no. 1, pp. 21–46, 1997.

[59] I. Sommerville, Software Engineering, Harlow, England: Pearson Education Limited, 2000.

[60] M. Z. Gül, "YazilimGeliştirmeSürecininİyileştirilmesi Ve TürkiyeUygulamalari", İstanbul Teknik Üniversitesi Fen BilimleriEnstitüsü, İstanbul, 2006.

[61] S. E. ŞEKER, "YazılımGeliştirmeModelleri ve Sistem/YazılımYaşamDöngüsü", *YBS Ansiklopedisi,* cilt 2, no. 3, 2015.

[62] O. Doğan, S. Bitim and K. Hızıroğlu , "A V-Model Software Development Application for Sustainable and Smart Campus Analytics Domain", Sakarya University Journal of Computer and Information Sciences, Nis. 2021.

[63] K. Hızıroğlu, D. T. Eliiyi, O. Doğan, S. Bitim, E. Akkol, Z. N. Köstepen, A. O. Yıldırım, Y. Demir and B. Akça, Sürdürülebilir ve Veri AnalitiğiOdakliKampüsProjesiFizibiliteRaporu, Nov, 2020. Accessed on: Aug. 7, 2021. [Online]. Available: https://akilliuniversite.bakircay.edu.tr/Sayfalar/2517/fizibilite-raporu.

[64] D. M. Sönmez, "Mühendis Ve MühendisYardimcilarininYaşamBoyuÖğrenimindeMeslek", *Electronic Journal of Vocational Colleges,* pp. 1–7, 2011.

[65] K. Petri, L. Maarit, How to steer an embedded software project: tactics for selecting the software process model. *Information and software technology*, 2005, 47.9: 587–608.

[66] M. Jain, N. Kaushik and K. Jayavel, "Building automation and energy control using IoT - Smart campus," in *2nd International Conference on Computing and Communications Technologies*, 2017.

[67] S. Rawal, "IOT based Smart Irrigation System," *International Journal of Computer Applications,* pp. 7–11, 2017.

[68] O. Chieochan, A. Saokaew and E. Boonchieng, "IOT for smart farm: A case study of the Lingzhi mushroom farm at Maejo University," in *Computer Science and Software Engineering*, 2017.

[69] F. Folianto, Y. S. Low and W. L. Yeow, "Smartbin: Smart waste management system," in *IEEE Tenth International Conference on Intelligent Sensors, Sensor Networks and Information Processing*, 2015.

2

Performance Analysis of Deep Learning Models for Re-identification of a Person in a Public Surveillance System

Harsha Saxena and Leena Ragha

Ramrao Adik Institute of Technology, India
Email: harsha.saxena@rait.ac.in; hodce@rait.ac.in

Abstract

Nowadays, a huge network of cameras is deployed in public locations, generating enormous amounts of video data. These data are monitored manually and can only be accessed if the necessity arises to verify the facts. Automating the system can increase the quality of monitoring and can be beneficial for high-level surveillance activities such as person recognition, suspicious activity detection, and unwanted event prediction for timely notifications. Person re-identification (ReID) tries to identify and track the activities of persons across multiple cameras. The common challenges faced are non-overlapping cameras, occlusion, large changes of viewpoint, shadows, various scales, cluttered background, pose, and illumination across different fields of view.

This chapter proposes an automatic real-world video surveillance system model that can track and re-identify multiple persons from a single-camera tracking environment. The proposed system compares the effect of different benchmarked deep neural networks on MARS and iLIDS-VID video re-identification dataset. DenseNet121, SEResNeXt50, Resnet50M, Resnet-50, and inceptionresnetV2 are the well-known neural networks used in the various image or video analytics domains. We train these models to analyze the effect in the re-identification domain. Further, we also analyzed the model based on the effect of hyper-parameters like learning rate, number of epochs, dropout rate, loss function, and pooling strategies. The results obtained by the training DenseNet121 model using the MARS dataset are 66.0% and the iLIDS-VID Dataset is 76.2%. The system's benefit is that it does not require a pre-stored database of individuals in advance at the time of testing for person recognition; it will create the dataset in real time. It will be also beneficial for crime control and prevention. Based on the experimental results produced, we conclude that standard deep learning techniques did not perform well due to various challenges including computational requirements, adequate training, and environmental issues. Hence, there is a need to investigate a suitable solution for a real-world automatic surveillance system.

2.1 Introduction

Surveillance cameras are affordable and are nowadays used everywhere, but the workforce they take to monitor and evaluate is costly. Therefore, the camera's recordings are usually sparingly checked or not checked in any way; they also are used solely as an index, to look back to when an incident has happened. Surveillance cameras can be a much more effective method if they are used to detect activities that need intervention in real time instead of capturing passively videos. This is the purpose of automatic visual surveillance: to acquire a summary of what is happening in the monitored domain, and then, based on this understanding, to implement effective actions. Visual monitoring for humans is one of the most active computer vision research subjects.

The need for security surveillance systems is significant in the circumstances in which human well-being could be compromised because of criminal activity. According to the Indian National Crime Records Bureau (NCRB) information in 2018 [1], the total crimes were reported to be 5.07 million comprising 3.13 million Indian Penal Code (IPC) crimes and 1.94 million Special and Local Laws (SLL) crimes, showing an increase of 1.3% over 2017 (5.007 million crimes). During 2018, IPC crimes have increased by 2.3%, and SLL crimes have declined by 0.1% over 2017. Crime against the human body, women, children, property, public order, and others come under IPC crime. Gambling, Explosives and Explosive Substance Act, Forest Act, Child Marriage Restraint Act, Arms Act, Terrorist and Disruptive Act, and others come under SLL crime. According to a survey, a total of 1,040,046 cases of offenses affecting the human body were reported, which accounted for 33.2% of total IPC crimes during 2018 as shown in Figure 2.1.

Security and crime control concerns are the motivating factors for the deployment of intelligent video surveillance systems that can emphasize the important facts and filter out routine conditions that do not represent a threat to security. To develop such intelligent systems, it is essential to identify and recognize a person(s) with high accuracy through a real-time multi-camera

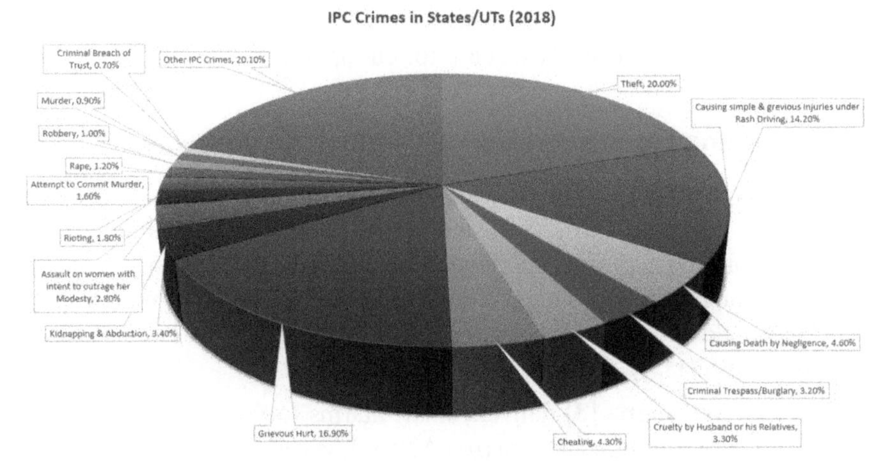

Figure 2.1 IPC crime rate in 2018 [1].

environment under diverse conditions to prevent/reduce the effect of crime on time. Currently, visual surveillance in dynamic scenes attempts to detect, recognize, and track certain objects from image sequences from multiple cameras. It is still difficult to create an intelligent surveillance system since there are a lot of difficulties that need to be solved. The main concern is how such a machine will comprehend human behavior. Despite substantial progress in human motion reconstruction, visual re-identification and comprehension of human activity or behavior are still in their infancy. It is hard to model a field based on real-life scenarios as not much work has been done in this field. The objective of visual surveillance is not only to place cameras in the position of human eyes but also to automate as much of the monitoring work as feasible. To improve a smart surveillance system, our study will focus on the development of a person re-identification framework.

This chapter is organized as follows. Section 2.2 discusses the currently present techniques and analysis of previous research associated with our proposed methodology. Section 2.3 presents models of computation and theoretical tools that we have adopted to re-identify persons. Experimental results and also the comparative analysis of varied algorithms are shown in Section 2.4. Section 2.5 concludes the outcomes obtained by using our work.

2.2 Literature Survey

This section introduces the currently existing techniques, tools, commercial products, and an analysis of previous state-of-the-art research papers on person re-identification using multi-camera tracking. It will also discuss the methodologies developed by the researcher along with the dataset used. The related research is described as a base for our approach.

2.2.1 Existing video surveillance commercial products

i. **Smart Things [2]:** The Smart Things local mobile application enables clients to control, robotize, and screen their home conditions using a mobile device. The application is arranged to meet every client's requirement. Smart Things compatible devices include motion, presence, and moisture sensors to control locks, electrical outlets, garage door openers, speakers, and thermostats.

ii. **Honeywell home [3]** delivers smart alarms to your cell phone based on face recognition and Amazon Alexa. It is easy to set up and operate, and it grows with your needs, thanks to the wide range of sensors and services available. Setup and usage are easy, with the ability to switch settings organically as you travel every which way, tracking and capturing moving objects as well as sounds. You will get an alert on your mobile device if it detects someone it does not recognize, and you will be able to see a video of the person. It notifies that the children are home, incited by facial recognition.

iii. **Google Home device [4]**, works with Nest Thermostat, Smart-Things, PhilipsHue, WeMo, and a couple of other arbitrary gadgets. Google Home uses your voice to bolt your entryways, arm or incapacitate your security framework, change your indoor regulator, and more.

All the commercial products available in the market operate electronic appliances, lighting, offer verbal directions, or sound an alarm when an intruder is detected. People's lives are made simpler and safer because of these products. Despite this, the intruder's action is done passively by watching the recorded video footage. So, there is a need to automate these products for suspicious activity detection and alarming. The commercial products mentioned above will not operate in public areas as efficiently as it does at home, because of the crowded unknown environment. We need a system that can automatically identify people and detect suspicious behavior in a scene in any public place.

2.2.2 A general automated visual surveillance system framework

An automated visual surveillance system's processing architecture consists of the following steps [5], as depicted in Figure 2.2: object detection, object

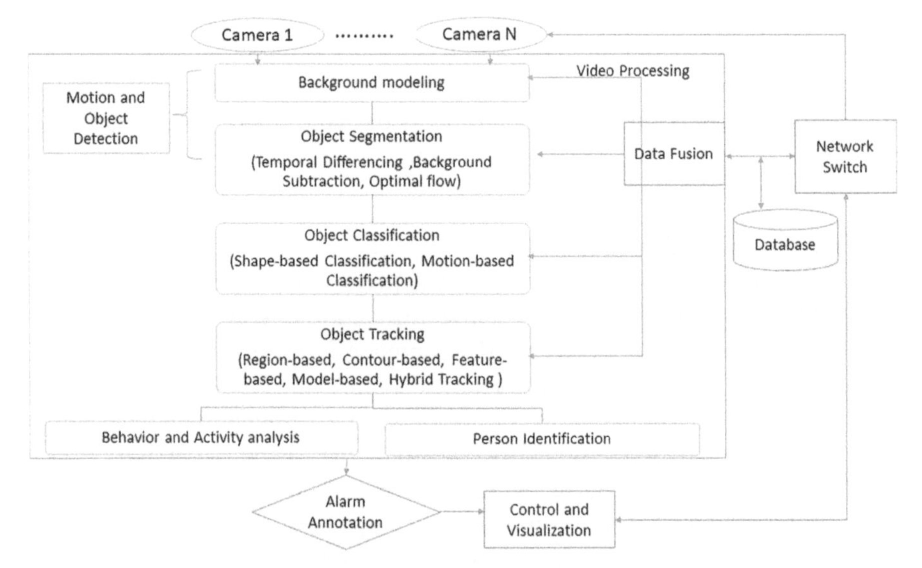

Figure 2.2 A general automated visual surveillance system framework [5].

classification, object tracking, behavior and activity analysis and understanding, person identification, as well as data fusion and camera handoff.

i. **The acquisition module:**
In the visual surveillance market, there are three most widely used visual surveillance technologies: thermal cameras, night vision devices, and CCD cameras.

ii. **Motion and object detection:**
Methods of motion detection aim to identify related areas of the pixels representing moving objects within the scene. Environmental (background) modeling and motion segmentation typically engage in the mechanism of motion and event recognition. Segmentation techniques are usually used in temporal or spatial details in the image series. Such widely used motion segmentation approaches include temporal differencing, background subtraction, and optical flow.

iii. **Object classification:**
In natural scenes, different moving regions may correspond to different moving objects. It is essential to correctly classify moving objects, to further track objects and analyze their behavior. There are two principal classes of methodologies for classifying moving objects: motion-based classification and shape-based classification.

iv. Object tracking:

In the object tracking module, cameras are used to identify and track moving objects, which are then converted into 3D world coordinates. Feature-based tracking, region-based tracking, model-based tracking, contour-based tracking, and hybrid tracking are the primary categories of tracking technologies.

v. Behavior analysis and understanding:

An automated visual surveillance system typically necessitates a dependable mix of image processing and artificial intelligence algorithms for semantic behavior learning and comprehension from video surveillance operations. Low-level image features are provided by image processing techniques. Expert decisions are provided by artificial intelligence techniques. To recognize scenarios, the computer vision community uses a probabilistic or neural network approach. To recognize scenarios, the artificial intelligence community uses a symbolic network. Researchers are also interested in unsupervised behavior learning and recognition, which uses a visual interpretation system to learn and recognize the most common scenarios in a scene without requiring the user to define the behavior beforehand for the system to work. For automated understanding and interpretation of the video, different types of concepts are required: basic properties, states, events, and scenarios.

vi. Person identification:

In the majority of video surveillance system literature, motion analysis and matching are used for the person identification, such as gait, posture, gesture, and comparison. Human gait and face are presently viewed as the fundamental bio-metric highlights that can be utilized for personal identification in visual surveillance systems.

vii. Camera handoff and data fusion:

To extend the surveillance area and give various view data to re-identify a person, the vast of visual surveillance systems are multiple camera-based. The multi-camera system should be able to track objects and identify their behavior predefined by a series of events or situations with overlapping fields of view or even learn new activity patterns. Each camera administrator identifies and tracks objects in the scene, and the data returns to the collected server where linked and melded objects are monitored. This result is named for the video event detection module, which recognizes and analyzes spatial and temporal events involving objects. A single-camera tracking will quickly create errors due to occlusion or depth. This doubt can be overcome from another angle.

A large majority of today's automated video surveillance systems execute low-level operations with high accuracy. Person re-identification and human activity detection have been the focus of academic attention in recent years.

Current advancements and problems faced by automated surveillance systems to re-identify persons will be discussed in the below sections. Person re-identification (ReID) tries to identify and track the activities of persons within or across cameras. To recognize the same person when it appears in different cameras, ReID algorithms need to extract representative characteristics of the target. This gives us a vast variety of options for how we may use the technology in the future. Person re-identification can be done in a single or multiple camera environment, utilizing still or moving cameras constantly or over a long period. In this chapter, we investigate person re-identification using single or multiple cameras continuously with still cameras for public safety. Throughout this chapter, we will refer to person re-identification as ReID.

2.2.3 Multi-camera tracking and person re-identification

Multiple camera tracking and person re-identification (ReID) are closely linked concepts. Multi-camera tracking refers to the ability to determine a person's location based on video feeds captured by many cameras. Applications such as visual surveillance, suspicious activity identification, anomaly detection, sports player monitoring, and crowd behavior analysis are made possible by the multi-camera trajectory.

Multi-camera tracking faces a few challenges; to reduce expenses, cameras are often placed widely apart, and their fields of vision do not usually overlap. To achieve this, there must be long periods of occlusion and significant shifts in perspective and lighting across multiple fields of view [10] as shown in Figure 2.3.

Similarly, the number of individuals is rarely known in advance, and the amount of data to be processed is enormous.

2.2.4 ReID system framework

The processing framework of a ReID system includes the following stages: person detection, person tracking, segmentation, feature extraction, feature matching, and person ranking as shown in Figure 2.4.

Figure 2.3 Subject intrinsic variations and person re-identification challenges.

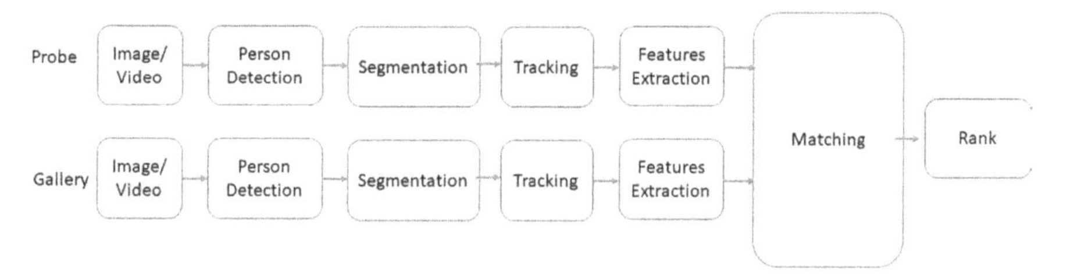

Figure 2.4 General framework for ReID system.

i. **Feature extraction:** The performance of ReID methods has improved by adopting new feature extraction techniques as shown in Table 2.1.

ii. **Metric learning:** This approach will be unsupervised and sub-optimal for ranking gallery candidates since it relies on a feature extraction technique paired with Euclidean distance. As a result of adding supervision based on training data to metric learning, the learning process becomes more efficient since the vectors of the same person are extremely similar, but those of other individuals are widely apart. Methods commonly used by the ReID community include the Euclidean distance, cosine distance, kernelize/non-kernelize Fisher discriminant analysis (FDA), RANK SVM, Keep-it-simple-and-straightforward metric (KISSME), kernelize/non-kernelize pairwise constrained component analysis (PCCA), etc.

iii. **Evaluation metrics:** Two metrics are used to evaluate person ReID accuracy that is mean average precision (mAP) and cumulative matching curve (CMC). Mean average precision is the mean across all queries' average precision. It is suitable for evaluating datasets in which an identity appears more than once in the gallery. The most common assessment methods for the re-identification of individuals are CMC [29]. Think of a basic single-gallery environment, with just one gallery

Table 2.1 Person re-identification feature extraction methods.

S. No.	Feature extraction	Year	Description
1	Ensemble of localized features (ELF) [15]	2008	It has two feature channels: color and texture. Color and texture histograms are generated for pedestrian detection and reacquisition using RGB, YCbCr, and HSV color spaces, and rotationally invariant Gabor and viewpoint invariant Schmid filters.
	Local descriptors encoded by Fisher vectors (LDFV) [16]	2012	Fisher vector representation encodes pixel position, intensity, and gradient information into a seven-dimensional local pixel descriptor.
3	IDECaffeNet [17]	2012	Using AlexNet architectures, each person image is considered as its class, and a convolutional neural network is trained for classification.
4	Dense color SIFT [18]	2013	Color histograms and SIFT features are retrieved from each patch based on unsupervised salience learning for addressing the viewpoint and pose variation in every image.
5	In gBiCov [19]	2014	Image representation (gBiCov) is based on a descriptor of covariance and multi-scale biologically inspired features (BIFs). This function is versatile for illumination, size, and context adjustments, making it perfect both for re-identification and for faces.
6	HistLBP [20]	014	There were four alternatives to ReID classification: regularized, pairwise constrained component analysis, kernel local Fisher discriminant analysis, marginal Fisher analysis, and a ranking collection voting scheme combined with varying sizes of histogram-dependent characteristics, as well as a linear and an RBF kernel.
7	IDE-VGGNet [21]	2015	A convolutional neural network is trained to classify large-scale images of each person.
8	IDE-ResNet [22]	2015	As CNN becomes deep, it is harder to train as gradients begin to vanish. This particular problem was addressed by ResNet by adding feature maps from the previous layers when skipping a layer and adding them to the next layer. Resnet50 is a 50-layer network with 3.8 billion FLOPS (floating-point operations per second).

(Continued)

Table 2.1 Continued

S. No.	Feature extraction	Year	Description
9	Local maximal occurrence (LOMO) [23]	2015	Model extract HSV color histograms and scale-invariant LBP features out of the image, which is then maximum pooled over the same horizontal strip to compensate for variations in perspective and light using the multi-scale Retinex technique.
10	WHOS [24]	2015	WHOS is a hybrid visual appearance descriptor for an image that extracts pose-invariant and illumination-invariant characteristics by splitting images into non-overlapping horizontal bonds and overlapping horizontal bonds.
11	GOG [25]	2016	Each strip of local patches in an image is given a Gaussian distribution model. We know that each strip contains Gaussian distributions, which are subsequently added together to form one Gaussian distribution.
12	Inception-v4 [26]	2016	Inception-v4 is a more homogeneous streamlined architecture with more modules than Inception-v3. The same input is manipulated by four different sub-networks, each with different learning abilities and depth. The Inception module, on the one hand, seems to improve our network's total depth. On the other hand, it expands the width which tends to improve the ability to learn. Inception-v4 is faster than ResNet.
13	Inception-ResNet-v2 [26]	2016	Inception network of residual connections, a Microsoft ResNet concept, with approximately the cost of computing Inceptionv4. Inception-ResNet-v2 was trained much quicker and achieved marginally higher final accuracy than Inception-v4, but the residual variant began to demonstrate inconsistency as the number of filters approached 1000.
14	DenseNet121 [27]	2017	DenseNet model improves gradient propagation by directly linking all layers. Here, instead of L connections, we have $L(L + 1)/2$ connections on a network of L.

(Continued)

Table 2.1 Continued

S. No.	Feature extraction	Year	Description
15	Resnet50M [28]	2017	The activation of conv5 (Sketch-a-Net) and res5a and res5b (Resnet-50) as the mid-level representation for fusion. Because of the significant pose-variation between views, global average pooling on res5a and res5b is employed to remove the spatial information.
16	SEResNeXt50 [33, 34]	2018	ResNeXt is a ResNet-based architecture in which the second convolution layer of each bottleneck block is implemented using grouped convolution. Furthermore, after each non-identity branch of the residual block, a squeeze-and-excitation block is used.

identity. A CMC top-*k* is the shifted step-function for each query to identify all the gallery samples according to their distance from the query small to high. The final CMC curve is determined by averaging all requests with the shifted phase functions. *K*-rank accuracy is 1 where samples from top-*k* galleries provide the name of the query, or 0.

2.2.5 Overview of previous Work in ReID

Over the past years, ReID has achieved remarkable success in different aspects [11–14]. A brief introduction of existing ReID studies from the following methodology-driven perspectives is shown in Figure 2.5. Many methods have been proposed and have utilized various features and learning techniques in video analytics for ReID.

As this chapter concentrates on multi-camera system tracking, there are mainly two types of approaches. Associating information between cameras and then across them is the initial step in the process. The second is to take into account all input detections at once. As a general rule, all input detections may be seen as a graph to formulate a global tracking issue. How related two nodes (detected objects) influence edge weights between them in a network? Based on the global approach, Wenqian Liu [6] developed an offline two-camera multi-object tracking system. They follow a maximal optimization problem of generalized multi-clique to take into account all the correlations between frame and camera data. They collect appearance and dynamic motion comparisons to measure good similarities as the input of the

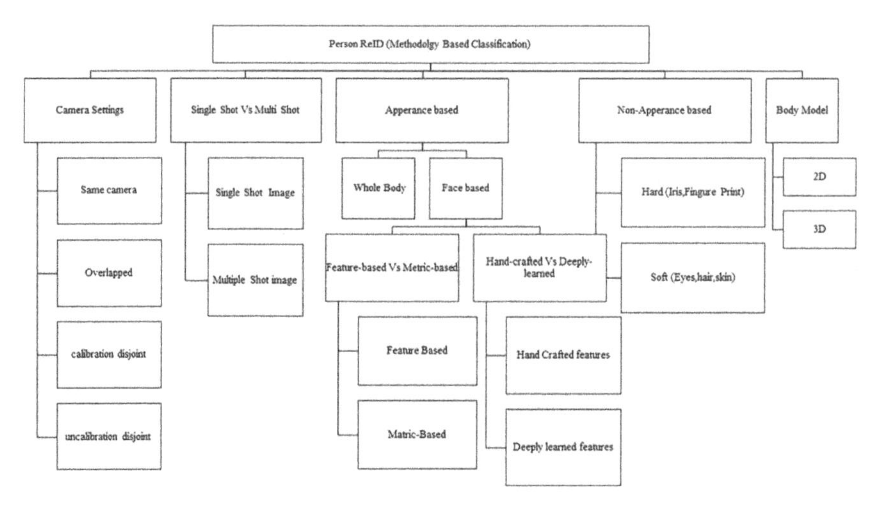

Figure 2.5 Person ReID (methodology-based classification).

graphic model. The local maximal occurrence representation (LOMO) function is used for extraction of a ReID, and Hankel matrix with iterative Hankel total least squares (IHTLS) algorithm for each tracklet target is performed for dynamic knowledge, and the rank estimates are used. They use terrace sequences from EPFL CVLAB and DukeMTMC dataset to evaluate their system. The algorithm still has space for improvements for tracking accuracy as the dataset has complex motion information and finding a better way for stitching the final track and merging the information got from GMMCP algorithm may be required.

To solve these problems, Ergys Ristani [7] has developed a new triple loss with real-time weights, in combination with a new hard identify mining methodology combining complicated and random identities, which achieves the best performance in both the MTMCT and ReID, if calculated using IDF1 [9], MOTA, or rank-1 scales.

Mingyue Yuan [8] combined a re-ranking and multi-image ReID task. The image pool is proposed for each identification to be stored. For a single image pool, they optimize sample diversity with our upgrade rules. Then multi-image system to re-ranking the target person. To test the performance of the proposed system, a new data collection called "indoor train station dataset (ITSD)" was developed to assess the performance of the proposed system using cumulated matching characteristics (CMC) or mean average precision (mAP). They achieve 81% accuracy on PRID-2011 dataset, 66% accuracy on iLIDS-VID dataset, 95.57% accuracy on Market-1501 dataset.

The survey showed that deep neural networks outperform hand-crafted feature methods.

2.2.6 Multi-camera tracking and person-re-identification datasets

The various publicly available datasets that are used in benchmark evaluation for person re-identification in multi-camera setting are shown in Table 2.2.

Table 2.2 provides a statistical summary of each dataset based on the number of persons in the cameras (IDS), duration of each video, number of cameras in the environment, calibration setting, scenarios considered for actors overlapping, blind spots, FPS rate taken, indoor/outdoor environment, resolution of images, and the year of the release of the dataset.

2.2.7 Challenges faced by ReID system

In earlier sections, we reviewed the conversion of video technology, the market growth of video surveillance technology, and the present focus on research activities in the fields of intelligent solutions to real issues in diverse scenes and domains. The actual gaps between users' expectations of smart and/or visible video surveillance and the fact that state-of-the-art technology solutions can be offered have yet to be bridged, even though the drive for political and commercial development and society-wide readiness to adopt this technology usually exists. Technologically, there are many research gaps yet to be resolved. Some may need a basic change in the thinking, design, and deployment of prior surveillance technologies to benefit from the latest development in components and the trend toward adapting technology to people. In the following, we list a few research gaps faced by automated surveillance systems to re-identify a person from multi-camera tracking.

- Multi-camera systems are applied to different environments with different configurations. As the number of cameras increases, the complexity of the system increases. It is very difficult to fuse data of different cameras in a large network. Human effort increases more when the network configuration changes. Therefore, there is a need for an automatic system for the calibration of cameras.
- The multi-camera system topology could be complex and difficult to calibrate as some camera views have overlapping and some have a disjoint view. It makes tracking of the person also challenging across the camera.

Table 2.2 Multi-camera tracking and person ReID datasets [9].

Dataset	IDs	Duration	Cam	Act or	Calib	Overlap	Blind spots	FP S	Resolution	Year
Indoor scenes										
Laboratory	3	2.5 min	4	Yes	Yes	Yes	No	25	320 × 240	2008
ApidisBasket	12	1 min	7	No	Yes	Yes	No	22	1600 × 1200	2008
NLPR MCT 3	14	4 min	4	Yes	No	Yes	Yes	25	320 × 240	2015
Outdoor scenes										
Campus	4	5.5 min	3	Yes	Yes	Yes	No	25	320 × 240	2008
Terrace	7	3.5 min	4	Yes	Yes	Yes	No	25	320 × 240	2008
Issia Soccer	25	2 min	6	No	Yes	Yes	No	25	1920 × 1080	2009
PETS2009	30	1 min	8	Yes	Yes	Yes	No	7	768 × 576	2009
USC Campus	146	25 min	3	No	No	No	Yes	30	852 × 480	2010
DukeMT MC	2834	85 min	8	Yes	Yes	Yes	Yes	60	1920 × 1080	2016
MARS	1261	N/A	6	Yes	Yes	Yes	Yes	25	1080 × 1920	2016
Mixed scenes										
Passageway	4	20 min	4	Yes	Yes	Yes	No	25	320 × 240	2011
Dana36	24	N/A	36	Yes	No	Yes	Yes	N/A	2048 × 1536	2012
NLPR MCT 1	235	20 min	3	No	No	No	Yes	20	320 × 240	2015
NLPR MCT 2	255	20 min	3	No	No	No	Yes	20	320 × 240	2015
NLPR MCT 4	49	25 min	5	Yes	No	Yes	Yes	25	320 × 240	2015
CamNeT	50	30 min	8	Yes	No	Yes	Yes	25	640 × 480	2015

- It will be difficult to detect and recognize a person across camera views because of the changes in viewpoints, illumination conditions, occlusion, crowded environment, and different camera views.
- ReID in a multi-camera environment needs unsupervised learning to avoid manual labeling of training samples and scenes. Existing systems used supervised learning and labeled the dataset manually.

This survey's significance stems from its attempt to address some of the major issues and shortcomings in the automated surveillance domain. Addressing the shortcomings could lead to the concept of extending the functionality of surveillance systems to become crime prevention tools.

For the person re-identification, we proposed to work using the DenseNet121, SEResNeXt50, Resnet50M, Resnet-50, and inceptionresnetV2 benchmarked deep neural networks. We apply these networks on the video Re-identification MARS and iLIDS-VID datasets which are created using real-world videos. Some of the above deep networks have been used to re-identify persons on image and video ReID datasets, but the results meet the real-world environmental issues. Although some of the selected networks are utilized in image and video analysis, they are not used in the field of re-identification. We are trying to determine the influence of these networks on person re-identification in real world.

2.3 Proposed System

The system design of the surveillance as shown in Figure 2.6 is proposed; it consists of multiple cameras that help to automatically detect and re-identify persons in public places. The development of this framework includes the following stages: background subtraction, person detection, person tracking, person verification in multiple cameras, and alarm generation.

i. **Dataset:** The datasets are usually manually labeled or detected by state-of-the-art person detection algorithms. It is independent of person detection. There are training set, query, and gallery. Training the model based on a training set and then extracting features from query and gallery, and calculating the similarity. For each query, return the top *N* similar images. In training and testing, there is no overlap of person IDs. In this chapter, the benchmark dataset used is motion analysis and re-identification set (MARS) and iLIDS video re-identification (iLIDS-VID) dataset.

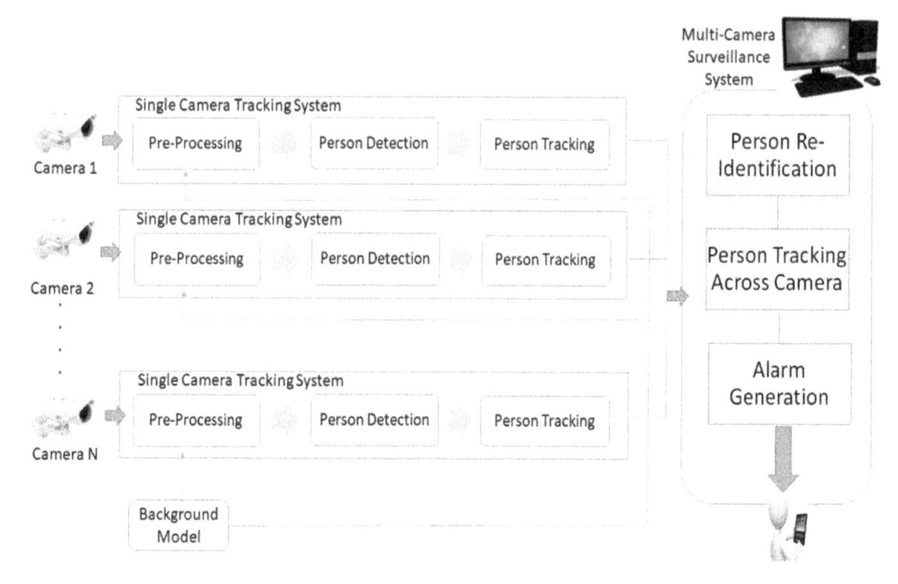

Figure 2.6 System design.

MARS (motion analysis and re-identification set) [35] is a sequence dataset, an expansion of the Market-1501 dataset. Six near-synchronized cameras on the campus of Tsinghua University gather 1200 videos, which are then pre-cropped to show the identified pedestrian in each video. On the other hand, there are sometimes some false positives included deliberately. One 640 × 480 SD camera and five 1080 × 1920 HD cameras made up the system. A total of 1261 pedestrians have been recorded by at least two cameras in MARS. Train and test sets are equally distributed across the 625 and 626 IDs in the MARS dataset. In general, the number of images per "tracklet" (a track fragment followed by a moving object) ranges from 2 to 920, with an average of 59.5 images.

iLIDS-VID (iLIDS video re-identification) dataset [36] is an image sequence dataset consisting of 300 identities, each of which comprises two picture sequences, for a total of 600. Due to environmental changes, this dataset is more difficult to analyze. A total of 150 identifiers are included in the iLIDS-VID dataset, which is evenly split. Each tracklet has a different number of pictures, ranging from 22 to 192, with an average of 70.8 images.

ii. **Pre-processing and person detection:** The video was captured by CCTV using the Python OpenCV package. Initially, the system performs pre-processing on frames by resizing the height and width and converting the image into a 416 × 416 RGB image. Person detection is a difficult problem that involves using methods to identify the human being (e.g., where he is), locating the person (e.g., how far his person exists), and classifying the individual with respect to other objects.

The proposed method uses "You Only Look Once" or the YOLOv3 [30–32] pre-trained model for the detection of a person.

iii. **Person tracking:** Individual person detected by the above module is fed into the DenseNet121 architecture to extract features of the person for re-identification. Then systems track a moving person frame-to-frame in an image sequence. Tracking involves the matching of features in consecutive frames.

iv. **Person re-identification:** Multi-camera tracking system helps to match the target and recognize it whenever the same target appears. The feature of a person is matched with all the person features that already exist in the database. Now if the person entered is the first person for the system, then he will be assigned with PersonID = 1. If he is not the first person, that means the system has few person features stored in the database; so now the system uses cosine distance similarity metric learning to find the similarity between two persons.

The database is an excel file that stores the features of every person. To increase the likelihood of matching the person with the same person, the system stores every fifth feature of the same person because as we are not considering only the face feature to match a person, we are considering the whole person appearance (color and HOG features). The person features get changed as he moves in a different angle or takes different postures; so saving all the features of the same person increases the likelihood of recognition and re-identification.

v. **Alarm generation:** A surveillance system is used for the security of humans. By combining the software and camera, we can use this system as an intelligent monitoring. After detecting a suspicious person, an alert message is sent to the owner's mobile.

2.4 Experimental Results and Discussion

Various experiments were carried out to realize the person re-identification system. First, we looked at person detection accuracy and worked on the

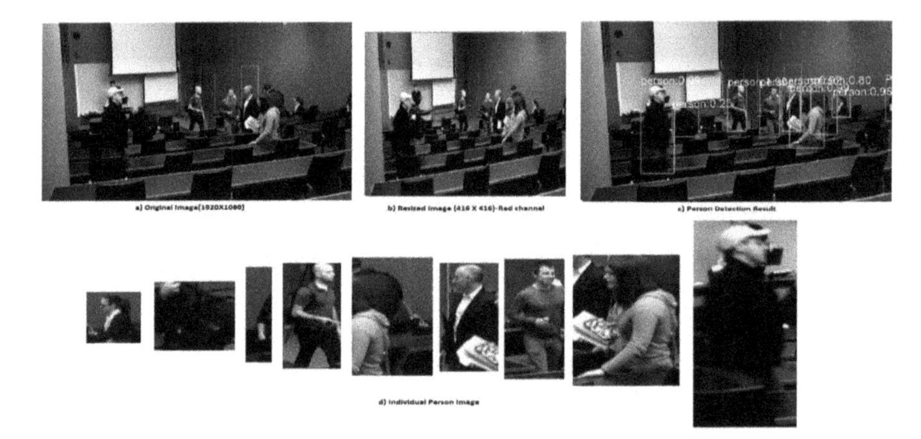

Figure 2.7 Results of pre-processing and person detection of Frame 640 from CAMPUS-Auditorium view-IP2.mp4 video.

pre-processing to improve the quality and size of an input image. We also decided on the thresholds for person detection. These results are presented in Section 2.4-i. We further worked with various benchmarked deep neural networks and analyzed them based on different hyper-parameters. These results are discussed in Section 2.4-ii. Further, Sections 2.4-iii and 2.4-iv discuss the results of ReID model training, testing of person feature matching, ReID, and tracking.

i. **Results of pre-processing and person detection:** The videos recorded by CCTV are accessed using Python OpenCV. Initially, the system performs pre-processing on frames by resizing the height and width and converting the image into a 416 × 416 RGB image. Then the person detection is performed using the YOLOv3 algorithm as shown in Figure 2.7, where each person is bounded using a bounding box and these bounding boxes are fed into deep neural network for person feature detection. For testing the model, CAMPUS real-world dataset [41] is used.

   In Frame 640 of Campus-Auditorium, as shown in Figure 2.8, it is observed that if the part of a person's body is also present in some part of the frame (shown with red color), those parts are also considered as a person. The proposed system ignored such cases in further processing by considering the only person with confidence greater than 0.85.

ii. **Effect of benchmark deep neural network and hyper-parameters on model training:** Different benchmark deep neural networks, as

Figure 2.8 Case of false person detection (in red color) of Frame 640 from CAMPUS-Auditorium view-IP2.mp4 video.

shown in Table 2.3, are trained to extract the person features for ReID on MARS and iLIDS-VID dataset. Two metrics are used to evaluate person ReID accuracy mean average precision (mAP) and CMC curve. The comparative analysis is based on accuracy as shown in Figures 2.9 and 2.10 and training loss as shown in Figures 2.11 and 2.12. It is shown that the DenseNet121 architecture provides a better result on MARS and iLIDS-VID datasets.

The effects of hyper-parameters like number of workers, height and width of an image, sequence length that is the number of images to sample in a tracklet, batch size, learning-rate, gamma, weight-decay, number of epochs, dropout rate, loss function, and pooling strategies are also analyzed on both datasets.

Because of hardware limitation, few hyper-parameters are kept fixed, such as the batch size is fixed to 8, number of workers to 4, gamma to 0.1, weight-decay to 5e-04, and sequence length to 4. But few parameters like learning rate, number of epochs, dropout rate, loss function, and pooling strategies are evaluated to see the effect on person re-identification. These results are discussed in detail in the following paragraphs.

- **Effect of learning rate on person re-identification result:**
  The learning rate is a hyper-parameter that controls how much the model is altered each time the model weights are changed in response to the predicted error. The choice of the rate of learning is complicated since a very low value will result in a lengthy training period that can be

Table 2.3 Results of ReID on MARS dataset.

S. No.	Backbone architecture	Model size	mAP	CMC curve (epochs = 100, learning rate = 0.0003, pool = avg. loss = hard_triplet_loss + cross_entropy_loss)				Time required (h:m:s)
				Rank-1	Rank-5	Rank-10	Rank-20	
MARS dataset								
1	DenseNet121	7.59448 M	62.3%	73.9%	87.4%	91.2%	93.7%	3:02:46
2	SEResNeXt50	26.79152 M	62.0%	75.6%	87.8%	91.0%	93.8%	4:11:20
3	Resnet50M	29.62603 M	59.9%	70.8%	85.0%	89.3%	92.7%	3:09:22
4	Resnet-50	24.78866 M	56.9%	69.3%	85.2%	89.4%	92.8%	3:03:40
5	inceptionresn etV2	55.26709 M	32.2%	47.6%	63.8%	71.1%	77.2%	4:42:02
iLIDS-VID Dataset								
6	DenseNet121	7.10761 M	77.5%	66.0%	91.3%	94.7%	97.3%	0:43:26
7	SEResNeXt50	25.81825 M	68.5%	55.3%	84.0%	92.7%	98.0%	1:00:51
8	Resnet50M	29.62603 M	79.9%	70.0%	92.7%	98.0%	98.7%	0:45:03
9	Resnet-50	23.81538 M	72.0%	60.7%	84.7%	91.3%	97.3%	0:44:29
10	inceptionresn etV2	54.53701 M	30.2%	15.3%	50.7%	62.7%	76.0%	1:23:44

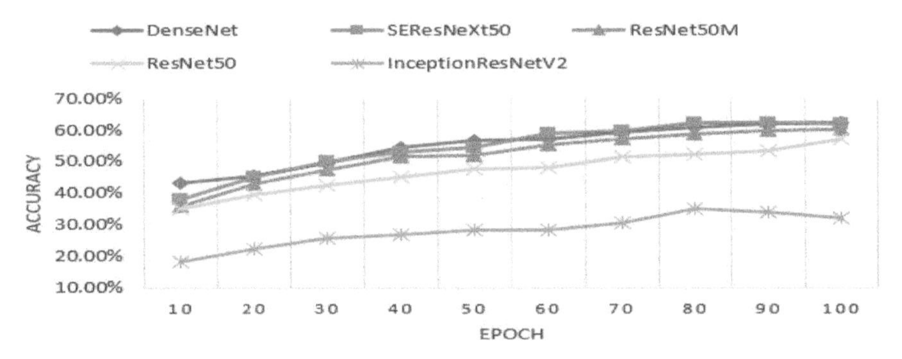

Figure 2.9 Accuracy vs. epoch graph on MARS dataset.

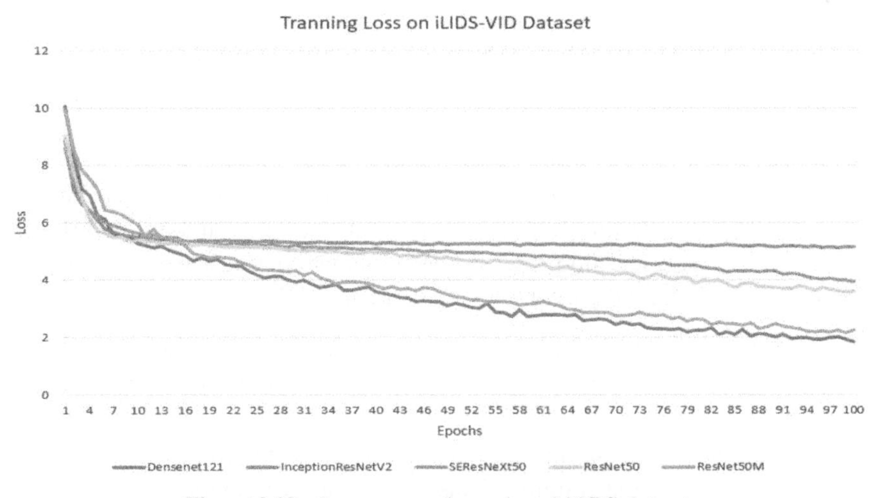

Figure 2.10 Loss vs. epoch graph on MARS dataset.

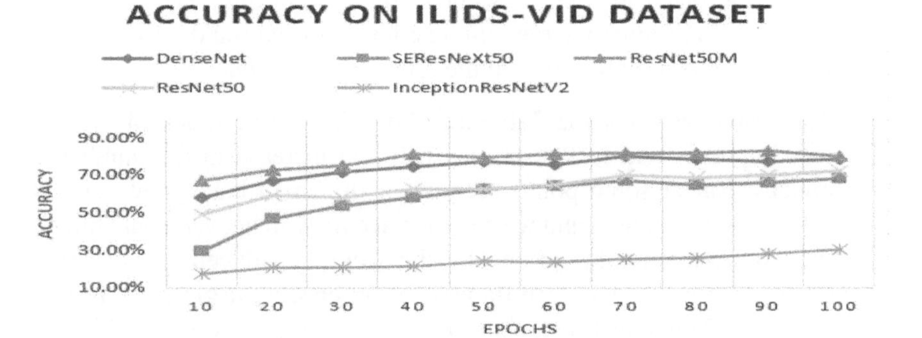

Figure 2.11 Accuracy vs. epoch graph on iLIDS-VID dataset.

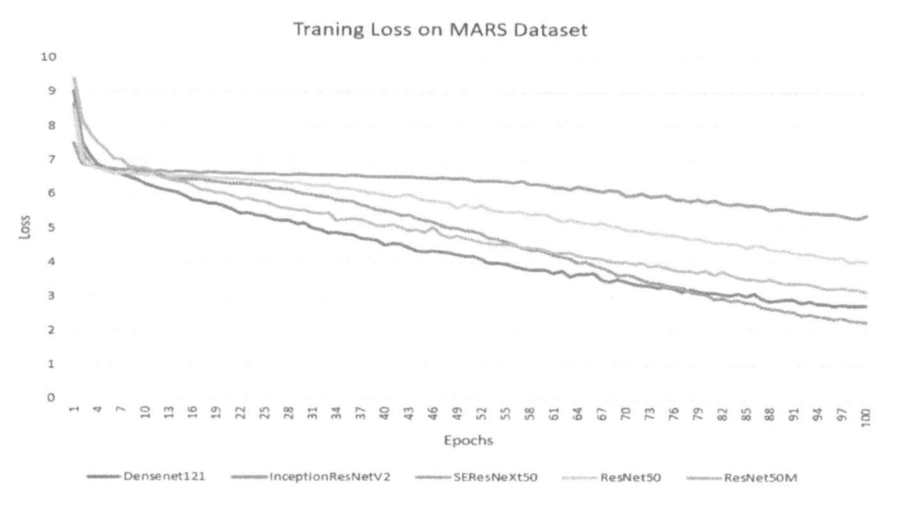

Figure 2.12 Loss vs. epoch graph on iLIDS-VID dataset.

stuck, whereas a very high value can lead to a too-quick or inconsistent training of a subset of weights. The results shown in Table 2.4 show that with the increase in learning rate, the accuracy decreases. But with the smaller learning rate, the system takes a large amount of time for training. The learning rate of 0.0003 shows the best result on MARS and iLIDS-VID datasets.

- **Effect of number of epochs on person re-identification result:** Epoch is a hyper-parameter that determines the number of cycles that the learning algorithm in the whole training dataset can run. One epoch means that each sample has been able to change the internal model parameters in the training dataset. One or two batches are part of an epoch. For the proposed method, an epoch that has eight batches is called the batch gradient descent learning algorithm. The number of epochs is traditionally large, often hundreds or thousands, allowing the learning algorithm to run until the error is minimized.

The results as shown in Table 2.5 show that the accuracy of the system improves and loss decreases with the increase in the number of epochs. But at some point, because of the small batch size network stuck in local minima and we get good accuracy in between but with the increase in the number of epochs, the accuracy reduced. For example, when the proposed system trained for 50 epochs, it gives an average loss of 4.321761, at 100 epochs, it gives an average loss of 2.752075, at 200 epochs, it gives an average loss of 1.819345, at 400 epochs, it gives

Table 2.4 Effect of learning rate on person re-identification.

S. No.	Learning rate	mAP	CMC curve (epochs = 100, pool = avg. loss = hard_triplet_loss + cross_entropy_loss) Rank-1	Rank-5	Rank-10	Rank-20	Time required (h:m:s)
DenseNet121 (MARS dataset)							
1	lr = 0.01	30.3%	45.2%	64.5%	71.9%	77.7%	2:57:47
2	lr = 0.001	61.3%	72.9%	86.7%	90.2%	93.5%	2:57:15
3	lr = 0.0001	55.1%	68.3%	84.0%	89.2%	92.7%	3:00:20
4	lr = 0.0002	59.9%	72.0%	86.4%	91.1%	93.7%	3:02:44
5	lr = 0.0003	62.3%	73.9%	87.4%	91.2%	93.7%	3:02:46
6	lr = 0.0004	62.1%	73.9%	87.1%	90.5%	93.4%	3:46:03
DenseNet121 (iLIDS-VID dataset)							
7	lr = 0.01	38.4%	25.3%	54.7%	65.3%	84.0%	0:46:52
8	lr = 0.001	78.0%	68.7%	90.0%	93.3%	96.7%	0:46:04
9	lr = 0.0001	72.6%	62.0%	84.7%	93.3%	96.0%	0:46:06
10	lr = 0.0002	75.9%	64.0%	90.0%	95.3%	97.3%	0:44:27
11	lr = 0.0003	77.5%	68.0%	90.7%	95.3%	99.3%	0:47:13
12	lr = 0.0004	76.9%	65.3%	92.0%	94.7%	97.3%	0:44:21

Table 2.5 Effect of number of epochs on person re-identification (MARS dataset) on DenseNet121 architecture.

S. No.	Epochs	mAP	CMC curve (learning rate = 0.0003, pool = avg. loss = hard_triplet_loss + cross_entropy_loss) Rank-1	Rank-5	Rank-10	Rank-20	Time required (h:m:s)
1	50	55.9%	69.3%	84.3%	88.7%	92.6%	2:11:51
2	100	62.3%	73.9%	87.4%	91.2%	93.7%	3:02:46
3	200	65.0%	76.0%	88.6%	92.4%	94.6%	6:03:42
4	300	65.1%	76.7%	88.7%	92.1%	94.6%	10:15:35
5	400	65.8%	76.4%	90.1%	93.2%	95.5%	11:40:08

an average loss of 1.555862, at 690 epochs, it gives an average loss of 1.532307, and at 800 epochs, it gives an average loss of 1.546234. So at 690 epochs, it gives the minimum loss and 66.6% mAP accuracy because of local minima or actual minima, but as the number of epochs increases the network comes out of local minimum. So, the number of epochs should be chosen wisely based on the accuracy achieved and the time taken for training the model.

Table 2.6 Effect of pooling strategy on person re-identification.

S. No.	Pooling strategy	mAP	CMC curve (epoch = 100, learning rate = 0.0003, loss = hard_triplet_loss + cross_entropy_loss)				Time required (h:m:s)
			Rank-1	Rank-5	Rank-10	Rank-20	
DenseNet121 (MARS dataset)							
1	Max	62.0%	73.3%	87.1%	91.0%	94.4%	2:57:02
2	Average	62.3%	73.9%	87.4%	91.2%	93.7%	3:02:46
DenseNet121 (iLIDS-VID dataset)							
3	Max	79.1%	69.3%	90.7%	94.7%	99.7%	0:46:38
4	Average	77.5%	68.0%	90.7%	95.3%	99.3%	0:47:13

- **Effect of pooling on person re-identification result:**
  Pooling is nothing but image sampling. The task of the pooling layer is to reduce the feature map resolution but to preserve the map characteristics needed in the classification by translation and rotational invariants. In addition to robustness and spatial invariance, pooling decreases computational costs dramatically. Back-propagation is used for pooling operations. It allows the processor to manage the process more efficiently.

  There are many different pooling methods used, such as max pooling, where the largest of the pixel values of a segment is considered, and average pooling, where average of the pixel values of a segment is considered. The results as shown in Table 2.6 shows that the average pooling gives better accuracy on MARS dataset in DesnseNet121 architecture over 100 epochs. Whereas, the maximum pooling gives better accuracy on iLIDS-VID dataset in DesnseNet121 architecture over 100 epochs.

- **Effect of loss function on person re-identification result:**
  To optimize the network, loss functions are used to estimate the loss of the model so that the weights can be updated to reduce the loss on the next evaluation.

  The proposed system used cross-entropy loss [37] and hard triplet loss [38] to optimize the network. The results as shown in Table 2.7 show that the combination of cross-entropy loss and hard triplet loss gives the

Table 2.7 Effect of loss function on person re-identification (MARS dataset) using DenseNet121 architecture.

			CMC curve (epoch = 100, learning rate = 0.0003, pool = avg)				Time required
S. No.	**Loss**	**mAP**	**Rank-1**	**Rank-5**	**Rank-10**	**Rank-20**	**(h:m:s)**
1	Hard triplet loss	50.9%	64.3%	82.7%	87.9%	91.3%	3:45:56
2	Cross-entropy loss	55.7%	68.4%	82.6%	86.4%	89.9%	2:50:22
3	Hard triplet loss + Cross-entropy loss	62.3%	73.9%	87.4%	91.2%	93.7%	3:02:46

better accuracy on MARS dataset in DesnseNet121 architecture over 100 epochs.

- **Effect of dropout on person re-identification result**
  The problem of overfitting and underfitting happens in machine learning and there are techniques to tackle these problems, which are called regularization techniques. While training, sometimes, models too might get overfitted and when during actual test, they may not probably perform well. Dropout regularization is one of the techniques used to tackle overfitting problems in deep learning. A dropout is an approach that, over the course of training, ignores randomly chosen neurons. They are randomly dropped out. This means that their contribution to downstream neuron activation is momentarily eliminated at the forward end and no weight updates are made to the backward neuron. Level deflation ranges between 0 and 1.

  The results as shown in Table 2.8 show that increasing the dropout rate will increase the accuracy of the system for a smaller dataset like iLIDS-VID by regularizing the effect of overfitting but giving the reverse effect on larger dataset like MARS. For a larger dataset, dropout will work for a larger number of epochs.

iii. **Results of ReID model training:** By considering the effect of hyper-parameters on model training, the proposed system has 0.0003 learning rate, average pooling for MARS dataset, and maximum pooling for iLIDS-VID, the combination of cross-entropy loss and hard triplet

Table 2.8 Effect of dropout rate on person re-identification.

S. No.	Dropout rate	mAP	CMC curve (epoch = 100, learning rate = 0.0003, pool = avg.) Rank-1	Rank-5	Rank-10	Rank-20	Time required (h:m:s)
DenseNet121 (MARS dataset)							
1	Dropout = 0	62.3%	73.9%	87.4%	91.2%	93.7%	3:02:46
2	Dropout = 0.2	57.9%	70.3%	85.5%	89.2%	92.6%	3:24:21
3	Dropout = 0.5	60.9%	73.3%	86.8%	90.8%	94.3%	3:02:44
4	Dropout = 0.8	59.7%	72.4%	86.7%	90.6%	93.3%	3:58:01
DenseNet121 (iLIDS-VID dataset)							
5	Dropout = 0	77.5%	66.0%	91.3%	94.7%	97.3%	0:43:26
6	Dropout = 0.2	75.4%	64.0%	88.7%	94.0%	96.7%	1:14:24
7	Dropout = 0.5	79.8%	70.0%	92.0%	96.0%	98.0%	0:43:18
8	Dropout = 0.8	80.6%	72.0%	92.7%	95.3%	98.0%	0:44:15

Table 2.9 Results of ReID model training on DenseNet121 Architecture.

S. No.	Dataset	Pooling strategy	Dropout rate	mAP	(Epoch = 800, learning rate = 0.0003, loss = hard_triplet_loss + cross_entropy_loss) Rank-1	Rank-5	Rank-10	Rank-20	Time required (h:m:s)
1	MARS	Avg.	0	66.0%	76.7%	89.8%	93.2%	95.4%	23:54:4 5
2	iLIDS-VID	Max	0.8	76.2%	65.3%	90.0%	94.7%	96.7%	5:54:17

loss function, no dropout rate for MARS dataset and 0.8 dropout rate for iLIDS-VID. These hyper-parameters are chosen with DenseNet121 architecture for getting the best accuracy on MARS and iLIDS-VID dataset as shown in Table 2.9. The training models are used to extract features of a person in real time.

iv. Results of person feature matching, ReID, and tracking: For feature matching system, use cosine distance metric learning. Now if the person entered is the first person for the system, then he will be assigned with PersonID = 1. If he is not the first person, then cosine distance between the query person and persons stored in the database is calculated. If the person is detected as a cosine distance less than 0.25 with some person, then he will be assigned with the ID of that person. And if the distance is greater than 0.25, then he will be considered as a new person and its feature gets stored in the database and assigned with the new ID. So, each person in the system will have a PersonID attached to it and will be tracked with the same ID as shown in Figure 2.13.

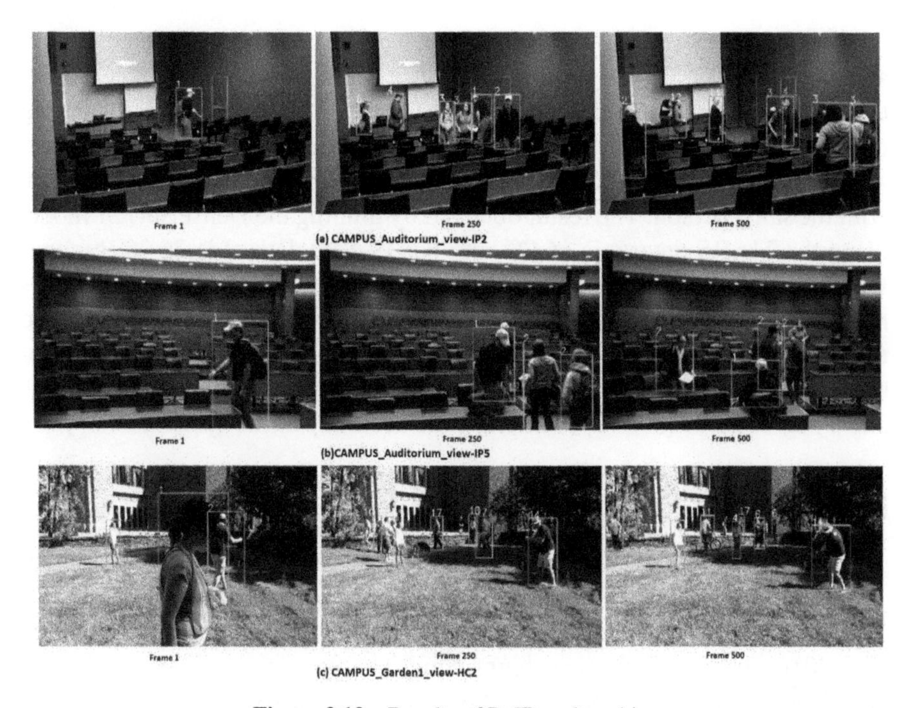

Figure 2.13 Results of ReID and tracking.

The time required to process a frame is more when new person features are stored in the database in comparison to the time required for matching the person. But as the number of people increases and the multiple features of the person are stored in the database, the number of comparisons increases and frames take a longer time to process. The database is an excel file that stores the features of every person. To increase the likelihood of matching the person with the same person, the system stores small deviated features of the same person because as we are not considering only face feature to match a person, we are considering the whole person appearance (color and HOG features). The person features get changed as he moves in a different angle or takes different postures. So saving all the features of the same person increases the likelihood of recognition and re-identification, but it increases the number of comparison and processing time. The accuracy of the system is not good, but it will get improved as the model accuracy improves with a proper framework that works in real time.

2.5 Conclusion

This work presents the person re-identification technique on closed circuit television cameras video using benchmark deep learning models which are trained on MARS and iLIDS-VID dataset. The proposed system detects a person using the YOLOv3 algorithm and the person features are extracted using trained DenseNet121 architecture. These features are used for ReID using cosine distance metric learning technique. Multiple features of the same person are stored in the database to increase the likelihood of matching the person with the same person. The result obtained by training the DenseNet121 model using the MARS dataset is 66.0% and iLIDS-VID dataset is 76.2% by considering all the effects of hyper-parameter variations. As the framework does not meet the real-time constraints, in the future, this framework is extended to multi-camera ReID by using advanced image enhancement techniques, feature extraction techniques, matching algorithms, hybrid deep neural networks, and a bag of tricks [39, 40] to improve the accuracy of the system.

References

[1] Crime in India by Statistical Branch of National Crime Records Bureau (December 2019) (Online)https://ncrb.gov.in/en/crime-india-2018

[2] Samsung SmartThings-Stay connected to your home and family (Online 2016). https://www.smartthings.com/

[3] Honeywell- Your connected home (Online 2016)http://homesecurity.honeywell.com/homeautomation.html

[4] Google- Get to know google home (Online 2016). https://madeby.google.com/home/

[5] T. Ko, "A survey on behavior analysis in video surveillance for homeland security applications," 2008 37th IEEE Applied Imagery Pattern RecognitionWorkshop, Washington DC, 2008, pp. 1–8

[6] Liu, Wenqian and Camps, Octavia and Sznaier, Mario. Multi-camera MultiObject Tracking. 2017

[7] Ristani, Ergys and Carlo Tomasi. "Features for Multi-target Multi-camera Tracking and Re-Identification". IEEE/CVF Conference on Computer Vision and Pattern Recognition, 2018:6036–6046.

[8] Mingyue Yuan, Dong Yin, Jinwen Ding, Zhipeng Zhou, Chengfeng Zhu, Rui Zhang, A Wang, A multi-image Joint Re-ranking framework with updateable Image Pool for person Re-Identification, Journal of Visual Communication and Image Representation, Volume 59,2019.

[9] Ristani, Ergys, Solera, Francesco, Zou, Roger, Cucchiara, Rita and Tomasi, Carlo. Performance Measures and a Data Set for Multi-Target, Multi-Camera Tracking, 2016

[10] Gong, Shaogang Cristani, Marco Yan, Shuicheng Loy, Chen Change."Person ReIdentification", Book 2013.

[11] Leng, Qingming Ye, Mang Tian, Qi. A Survey of Open-World Person ReIdentification. IEEE Transactions on Circuits and Systems for Video Technology.PP. 1-1.10.110/TCSVT.2019.2898940.

[12] Liang Zheng and Yi Yangand Alexander G. Hauptmann, "Person Reidentification: Past, Present and Future", CoRR,abs/1610.02984, 2016

[13] Khawar Islam, Person search: new paradigm of person Re-Identification: A survey and outlook of recent works, Image and Vision Computing, Volume 101, 103970, ISSN 0262-8856, 2020

[14] S. Karanam, M. Gou, Z. Wu, A. Rates-Borras, O. Camps and R. J. Radke, "A Systematic Evaluation and Benchmark for Person Re-Identification: Features, Metrics, and Datasets," in IEEE Transactions on Pattern Analysis and Machine Intelligence, vol. 41, no. 3, pp. 523–536, 1 March 2019, doi: 10.1109/TPAMI.2018.2807450.

[15] D. Gray and H. Tao, "Viewpoint invariant pedestrian recognition with an ensemble of localized features," in ECCV,2008.

[16] B. Ma, Y. Su, and F. Jurie, "Local descriptors encoded by fisher vectors for person Re-Identification," in ECCV Workshops, 2012

[17] Krizhevsky, I. Sutskever, and G. E. Hinton, "Imagenet classification with deep convolutional neural networks," in NIPS, 2012

[18] R. Zhao, W. Ouyang, and X. Wang, "Unsupervised salience learning for personRe-Identification," in CVPR, 2013

[19] Bingpeng Ma, Yu Su, Fr´ed´eric Jurie. Covariance Descriptor based on Bio-inspired Features for Person Re-Identification and Face Verification. Image and Vision Computing, Elsevier, 2014, 32 (6–7), pp.379–390. ff10.1016/j.imavis.2014.04.002ff. ffhal01009958

[20] F. Xiong et al., "Person Re-Identification using kernel-based metric learning methods," in ECCV, 2014.

[21] K. Simonyan and A. Zisserman, "Very deep convolutional networks for large-scale image recognition," arXiv preprint arXiv:1409.1556,2015

[22] K. He et al., "Deep residual learning for image recognition," arXiv preprint arXiv:1512.03385, 2015

[23] S. Liao et al., "Person Re-Identification by local maximal occurrence representation and metric learning," in CVPR, 2015

[24] G. Lisanti et al., "Person Re-Identification by iterative re-weighted sparse ranking" -PAMI, vol. 37, no. 8, pp. 1629–1642,2015.

[25] T. Matsukawa et al., "Hierarchical gaussian descriptor for person Re-Identification", in CVPR, 2016.

[26] Szegedy, Christian Ioffe, Sergey Vanhoucke, Vincent Alemi, Alexander. (2016). Inception-v4, Inception-ResNet and the Impact of Residual Connections on Learning. AAAI Conference on ArtificialIntelligence.

[27] G. Huang, Z. Liu, L. Van Der Maaten and K. Q. Weinberger," Densely Connected Convolutional Networks," IEEE Conference on Computer Vision and Pattern Recognition (CVPR), Honolulu, HI, 2017, pp. 2261–2269, doi:10.1109/CVPR.2017.243, 2017

[28] Yu, Qian Chang, Xiaobin Song, Yi-Zhe Xiang, Tao Hospedales, Timothy. (2017). The Devil is in the Middle: Exploiting Mid-level Representations for Cross-Domain Instance Matching.

[29] R. M. Bolle, J. H. Connell, S. Pankanti, N. K. Ratha and A. W. Senior," The relation between the ROC curve and the CMC," Fourth IEEE Workshop on Automatic Identification Advanced Technologies (AutoID'05), Buffalo, NY, USA, 2005, pp. 15–20, doi:10.1109/AUTOID.2005.48.

[30] Ayoosh Kathuria, URLhttps://towardsdatascience.com/yolo-v3-object-detection53fb7d3bfe6b

[31] Redmon, Joseph Farhadi, Ali. YOLOv3: An Incremental Improvement. ArXiv, abs/1804.02767, 2018

[32] J. Redmon, S. Divvala, R. Girshick and A. Farhadi," You Only Look Once: Unified, Real-Time Object Detection," 2016 IEEE Conference on Computer Vision and Pattern Recognition (CVPR), Las Vegas, NV, 2016, pp. 779-788, doi:10.1109/CVPR.2016.91

[33] J. Hu, L. Shen and G. Sun," Squeeze-and-Excitation Networks," IEEE/CVF Conference on Computer Vision and Pattern Recognition, Salt Lake City, UT, 2018, pp. 7132–7141, doi:10.1109/CVPR.2018.00745, 2018

[34] S. Xie, R. Girshick, P. Doll´ar, Z. Tu and K. He," Aggregated Residual Transformations for Deep Neural Networks," IEEE Conference on Computer Vision and Pattern Recognition (CVPR), Honolulu, HI, 2017, pp. 5987–5995

[35] Zheng, Liang Bie, Zhi Sun, Yifan Wang, Jingdong Su, Chi Wang, Shengjin Tian, Qi. (2016). MARS: A Video Benchmark for Large-Scale Person Re-Identification. 9910.868-884.

[36] Person Re-Identification by Video Ranking. T. Wang, S. Gong, X. Zhu and S. Wang. In Proc. European Conference on Computer Vision (ECCV), Zurich, Switzerland, September 2014

[37] C. Szegedy, V. Vanhoucke, S. Ioffe, J. Shlens and Z. Wojna," Rethinking the Inception Architecture for Computer Vision," 2016 IEEE Conference on Computer Vision and Pattern Recognition (CVPR), Las Vegas, NV, 2016, pp. 2818–2826, doi:10.1109/CVPR.2016.308.

[38] Hermans, Alexander Beyer, Lucas Leibe, Bastian. In Defense of the Triplet Loss for Person Re-Identification, 2017

[39] Pathak, Priyank Eshratifar, Amir Erfan Gormish, Michael. Video Person Re-ID: Fantastic Techniques and Where to Find Them (Student Abstract).

Proceedings of the AAAI Conference on Artificial Intelligence. 34. 13893–13894, 2020

[40] Porrello, Angelo Bergamini, Luca Calderara, Simone. Robust ReIdentification by Multiple Views Knowledge Distillation,2 020

[41] CAMPUS dataset for multi-view object tracking https://bitbucket.org/merayxu/multiviewobject-tracking-dataset/src/master/CAMPUS/

3

Exploiting Trajectory Data to Improve Smart City Services

LNC K. Prakash[1], G. Suryanarayana[2], and M. A. Jabbar[3]

[1]Associate Professor, CSE, CVR College of Engineering, India
[2]Associate Professor, CSE, Vardhaman College of Engineering, India
[3]Professor & HoD, CSE (AI&ML), Vardhaman College of Engineering, India

Abstract

Urban computing would be the method of acquiring, integrating, and analyzing large amounts of non-homogeneous information collected in urban areas by a number of sources which include sensing devices, gadgets, automobiles, housing developments, and humans in order to address significant problems such as environmental pollution, steadily increasing energy usage, as well as traffic problems. Urban computing links inconspicuous and ubiquitous sensor technology, powerful data management, and analytics techniques along with visualization methodologies. This allows for the collection of vast volumes of data related to the objects that are moving which called trajectories are providing potentially useful knowledge more about objects moving. Automated methods for understanding this material are known as trajectory data mining. The goal of this article is to analyze (1) the way in which trajectory data mining activities are being described at an existential level, (2) the exact sort of knowledge that can be extracted through trajectory data, and (3) the methods that trajectory data mining techniques apply toward various tasks of urban computing.

3.1 Introduction

The advancement of knowledge and communication advanced technologies, particularly in sensing system and wireless connectivity, is overflowing in data which contains time-varying phenomenological analysis. Even though this type of data is related to issues including expending our storage space as well as transmitting data frequency band, investigations have indicated that these data sources are a valuable resource. Their research will lead to strategies to significant research problems in a variety of fields, including urban planning, public transit, behavioral environmental sciences, athletics scene analysis, monitoring, and safety. By inventing methodologies and tools to handle application challenges, scholars in trajectory prediction have made vital contributions. There is an essential need for an interconnected picture of the technical challenges answered by utilizing trajectory data as well as the analysis techniques accessible to tackle every category of application-oriented problems [1] as investigators begin to discuss varied application challenges. It allows researchers to quickly identify difficulties that have already been handled, as well as technological gaps, and to derive recommendations for future formulation development from the relationships between challenges and techniques.

With this article, we hope that we can provide an inclusive approach of trajectory mining implementation challenges with respect to urban computing. An overview of programs that use the answers to these application difficulties is also provided, and calculative, quantitative, and graphical analytical techniques are all available. Computer-based approaches of trajectory mining are the subject of this article. The main goal of studying the trajectory data is to execute the process of urban computing. Because of urbanization's quick advancement, there are now a lot of big urban centers that have modernized a person's life, but they have also created a lot of problems, such as carbon emissions. As cities have become increasingly diverse and complex, trying to tackle these obstacles seemed somewhat unimaginable a few years ago. As a result of smart sensors and huge computational infrastructural facilities, urban neighborhoods are now awash in a diverse range of big data (e.g., human transportation, pollution levels, traffic conditions, etc.). As a result of this, big data provides a wealth of information about a city and, when using city-wide data on human transportation, we can, for example, identify the trouble spots in a city's road infrastructure. As a result of these findings, urban planning throughout the coming years can be done properly [14]. Some other instance is to identify the root causes of air pollution in urban areas by analyzing the significant relation between pollution levels as well as other sources of information, such as traffic conditions and landmarks [15].

To take advantage of these opportunities, we set ambitious goals of urban computing, which also intends to unlock the potential of understanding from big and diverse trajectory data collected in urban areas as shown in

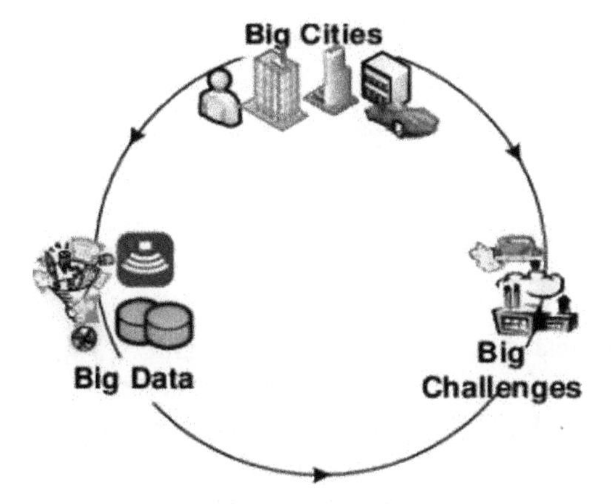

Figure 3.1 Motivation of urban computing.

Figure 3.1 and implement this potential knowledge to solve significant issues that our cities face daily. Using large amounts of trajectory data, we can handle the biggest problems in urban areas [16]. Using urban computing, urban areas can address big problems such as environmental pollution, enhanced power consumption, and traffic jams. It improves the effectiveness, incorporation, and analysis of big and heterogeneous trajectory data generated by different origins in urban landscapes, including sensors, equipment, automobiles, housing developments, and people.

Urban computing attaches unfussy and abundant smart sensors, sophisticated information management as well as analytic models, and novel optimization algorithms to generate bonus solutions that enhance urban landscape, living thing life quality, and town operations systems as shown in Figure 3.2. It also aids in the understanding of urban occurrences and the prediction of city's future by using urban computing techniques. Information technology in urban environments is a multidisciplinary field that combines computer technology science to applied areas like shipping and construction management as well as economics, ecosystems, and social science.

Continue reading to find out how this article is structured. A brief review of trajectory data and data mining for urban computing is as follows. A general framework of urban computing is described in Section 3.2. Section 3.3 describes trajectory data and trajectory data mining and different ways for mining trajectory data. Aside from that, Section 3.4 links applications of the trajectory data that take use of the answers to the application difficulties. Section 3.5 briefly summarizes open challenges that must be solved for trajectory data mining. Publicly available trajectory datasets are described in Section 3.6. Section 3.7 draws the conclusion and future works.

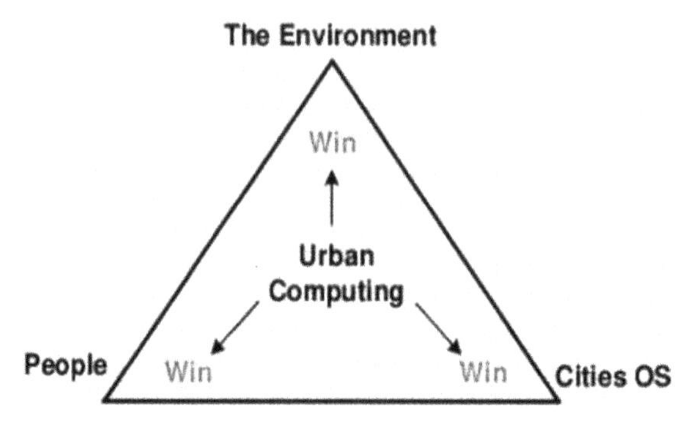

Figure 3.2 Objective of urban computing.

3.2 General Framework of Urban Computing

The general framework of urban computing has four layers, as shown in Figure 3.3, including urban sensing, urban database administration, data analytics, and service provisioning. Here is a quick rundown of how the structure works through using urban outlier detection as an example [17]. When it comes to monitoring people's mobility in sensing step (e.g., how they navigate a city's road network), we use GPS detectors or one's smartphone signals to continually review them. People's social media data are also constantly being collected. The data management step consists of putting together an indexing framework that integrates both spatial characteristics and text in addition to supporting effective data analytics. In the data analytics step, when an abnormality has occurred, we can describe the positions in which a person's scenarios are considerably different from their origin patterns using data analytics. In the service providing step, the positions and explanation of the irregularity will be forwarded to the concerning authority.

In this regard, some research challenges in urban computing are as follows:

- In what ways does urban computing pose a challenge to researchers?
- What are the research topic's hurdles?
- What are the most significant urban computing methods and techniques?
- What are some of the most application areas in this field, and how would an urban computation system function?

As a result of these concerns, we consider introducing urban computing as a proper term in this article, as well as its clear explanation and primary research issues. If the society can fully understand as well as analyze

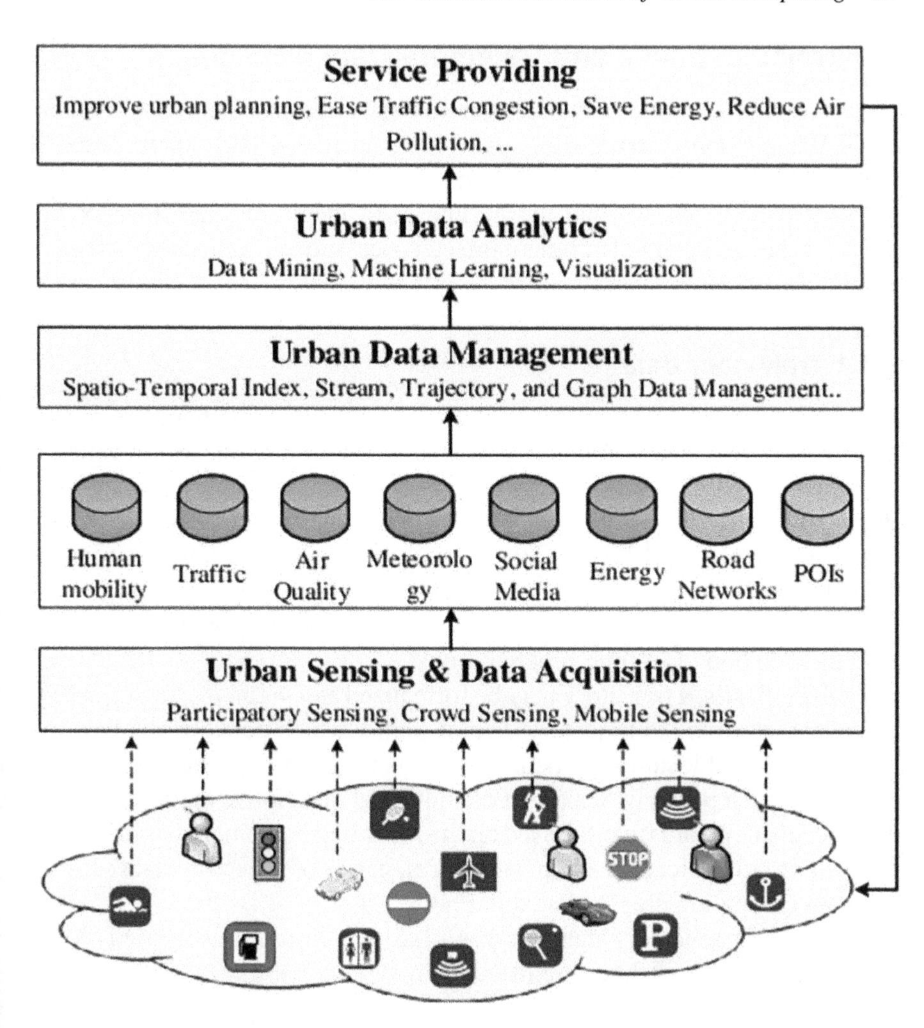

Figure 3.3 Framework of urban computing that uses trajectory data.

this transnational field, the reliably scientific studies and real systems could be generated to make cities that are more efficient and environmentally friendly. It is also a multidisciplinary scientific discipline in which computer science meets traditional city-related fields including structural engineering, transportation, finance, and power generation engineering as well as environmental studies, biodiversity, and social science from a computer science point of view. This document focuses primarily on the aforementioned challenges.

3.3 Trajectory Data and Trajectory Data Mining

Here, we provide a brief explanation of the data analysis theoretical foundations. When it comes to trajectory data for urban computing, experts describe it and situate it in the wider context of the larger field of data mining. Due to space constraints, we will not go into great length on these themes, but readers who are open to learning more can access to the sources presented in the article.

3.3.1 Trajectory data

Mobility information is accessible in various formats, depending on the method used to collect it. Global system for mobile communications, which is abbreviated as GSM, as well as geosocial network-based trajectory data were distinguished by Spinsanti *et al.* [1]. Radio-frequency identification which is simply known as RFID data as well as Wi-Fi based data were contributed by Pelekis [2] and Theodoris [3] are described about this data. Geographic coordinates are captured by a global positioning system's gadget conveyed by a moving body in a time-ordered series.

Specifically, a trajectory may be formalized as $T = (p_1, p_2, \ldots, p_n)$, where each characterizes the trajectory data of item k which contains the position identifier, spatial location of the position, the time at which the position was recorded, and a possibly empty list of additional descriptive data (e.g., direction, occupancy status, etc.). Based on the recording equipment, various ways can be used to depict a spatial position. Categories of trajectory data sources in urban computing are illustrated in Figure 3.4.

Different trajectory data sources [4–6] which are mostly used in the development of smart cities are listed below:

- There is a long history of people intentionally and casually capturing their travels in the spatial and temporal trajectories. For the aim of remembering a trip and communicating it with others, travelers record the GPS paths they take. Trails are recorded by cyclists and joggers to aid in game recaps as each snapshot is geotagged with a position and a time-stamp; a succession of geotagged photographs on Flickr may be used to create an imaginary geographical journey. Whenever ordered sequentially, a person's "verification" on a location-based social platform may be viewed as a trajectory which is referred to as active recording. On the other hand, passive recording refers to many geographical paths that are generated accidentally by smartphone owners who maintain a long list of cell tower identifiers with accompanying transition

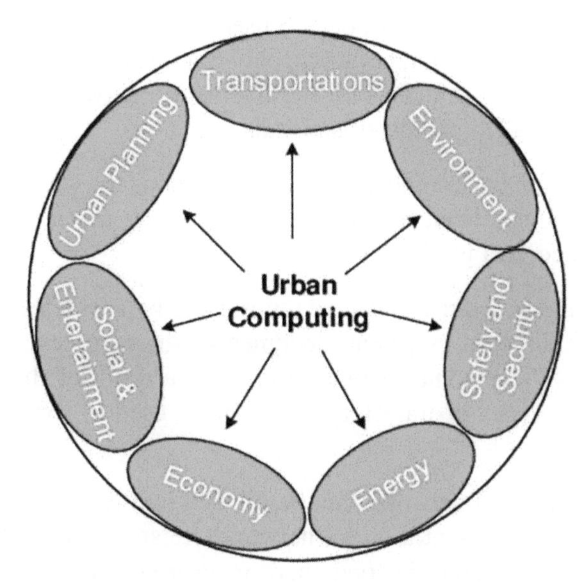

Figure 3.4 Categories of trajectory data sources in urban computing.

timings. As a result of the temporal stamp and vendor ID associated with each payment, a cardholder's geographical trajectory may also be determined from the card's event logs.

- A large percentage of GPS-equipped automobiles have emerged in the regular lifestyle (such as cabs, minibuses, ships, and airliners). A GPS sensor is installed in several cabs in urban areas, allowing vehicles to provide a time-stamped position on a regular basis. For allocation of resources [7, 8], traffic analysis [9–11], and upgrading transportation systems [12], these reports provide a significant number of geographical trajectories.
- Loop detectors, video surveillance, and flying cars are just a few methods to monetize traffic information. On expressways, loop sensors are typically paired together (e.g., highways). The duration between two successive (i.e., a pair of) sensors is recorded rather than the absolute time. A couple with traffic sensors separated by a certain range can be used to determine a highway's travel speed depending on an interval of time.
- The record of a phone call, including the caller as well as receiver's contact information, the call's launch time, and its period. We could also analyze the behavior of individuals or create a network among multiple

users if we had that kind of data. A user's commonality to another could also be determined. Other mobile phone signals are more involved with the user's location than with the interaction between mobiles. These signals are called geolocation signals.

- Insights from environmental monitoring among the weather reports that can be crawled from online sites are moisture, temperature, pressure, wind speed, and weather patterns.
- Financial characteristics can be represented by a wide range of sources (e.g., credit card payment documents, share prices, rental prices, and people's personal income levels). As a result of their cumulative use, these types of trajectory data can be used to forecast economic circumstances.
- Data from social networks can be divided into two categories. Graphs are used to represent a social network's relationships, interdependencies, and interactions. Second, there are user-generated social networks, such as text messages, pictures, and videos, which mostly includes a great deal of information about a user's habits and preferences. Due to the fact that we can design people's movement in urban areas with the existence of social media [18, 19], humans can specify interaction and recognize urban discrepancies.
- The energy demand of a city is reflected throughout the gas usage of automobiles on local roads and in service stations. A detector and other information sources can provide the information directly, while another source of data can provide it inherently (e.g., from the GPS trajectory of a vehicle). A city's electric grid (e.g., the dispersion of service stations) can be evaluated, and air pollutant emissions from automobiles on urban roads could be calculated, and the most fuel driving path can really be determined using the data. A building or an apartment complex power consumption can also serve as a tool for optimizing residential power use by trying to shift voltage levels to non-peak hours.
- Medical centers have amassed a vast amount of health services and disease information. Humans can also supervise their own health issues, including their heartbeat, pulse, and sleep cycles using their wearable smart devices. A cloud-based medical examination and diagnosis could be performed using the data collected. These types of trajectory datasets can be aggregated to examine the effect of climate changes on population lives as well as a person's health.

- Biologists have been tracking the movement patterns of animals such as tigers and birds to understand the animals' migration patterns and living conditions [13].
- Various natural occurrences are being tracked by meteorologists, environmentalists, climatologists, and oceanographers. Using these trajectories, scientists can better prepare for natural calamities and preserve the ecosystem in which we live.

Pre-processing of trajectory data:

Some primary methodologies that are the pre-processing of trajectory data before mining on it are introduced in this portion, which are noise filtering, stay point identification, trajectory contraction, and trajectory fragmentation.

- **Noise filtering:** Due to equipment noise and many other considerations, such as receiving poor locating signals in street environments, geographic trajectories will never be fully accurate. Certain types of errors, such as when a few GPS locations fall off from the actual route taken by a vehicle, a noise point may be too far away from its true location (e.g., several hundred meters) to provide helpful information, such as travel speed. Because of this, we need to eliminate any potential sources of noise before beginning a mining operation.
- **Detection of stay points:** Trajectories are not equally important from the perspective of spatial points. Shopping centers, tourist hotspots, and service stations where a vehicle was refueled are all examples of places that people have spent time. We call them "Stay Points." Sometimes, it is possible to turn a collection of time-stamped geographic locations into a constructive succession of locations using the stay points; therefore, it can be used for a wide range of applications, which include travel suggestions [20, 21], destination forecasting [22], and taxi recommendation [23]. That kind of stay points must be excluded from a trajectory during the preprocessing stage.
- **Compression of trajectory data:** For a moving object, researchers can essentially measure a time-stamp based on geography plan in advance each second. Aside from battery drain, this requires a significant amount of information exchange, computing, and data management. Many implementations do not require such a high level of position precision. To deal with this problem [24], two types of trajectory compression techniques (depending on trajectory structure) have been generally used, which try to minimize the volume of the trajectory without

affecting much accuracy in its new data illustration. As an example, there is offline compression (also known as batch mode), which diminishes a trajectory size once it has been produced in full. As an alternative, you can condense a trajectory as an entity travel, which is known as "online compression."

- **Segmentation of trajectory:** There are a number of scenarios where we need to segment a trajectory for further processing, such as clustering and classification. Not only does segmentation simplify computations, it also allows us to mine richer knowledge, such as sub-trajectory patterns that go beyond what we can learn from an entire trajectory. There are three main types of segmentation methods [25, 26] which are based on time interval, the shape of a trajectory, and the semantic meanings of points in a trajectory.
- **Map matching:** Latitude/longitude locations are converted to road segments using map similarity. To evaluate flow of traffic, guide the vehicle's tracking, anticipate where the vehicle is going, identify the most regular travel route between a source and a destination, etc. Knowing which highway a vehicle was/is on is essential. Due to parallel roads, overpasses, as well as spurs, map comparison is a difficult problem [27]. There are two ways to categorize map matching methodologies which are based on the additional details used or the number of sampling locations recognized in a trajectory.

Management of trajectory data:

Mining big trajectories takes a long time since we must repeatedly access various chunks of the trajectories or various portions of a trajectory. This necessitates the use of efficient data analysis approaches that allow for the retrieval of trajectories in a timely manner (or segments of a trajectory). In contrast to moving object collections, the trajectory data management described in this section is focused on the present position of a traveling item, whereas the trajectory data management outlined in the previous section is focused on the traveling record of an object moving. A lot more [28] provide a detailed survey of trajectory data handling.

- Trajectory query classification can be done through coordinate-based query which can be represented by Windows Query in which Select all objects within a given time slot and given period/slice, Nearest Neighbor Query and Approximate Query. Trajectory-based query which uses topological query in which some basic predicates are used and navigation query that uses information not directly kept in database but can be derived, for example, "At what speed does this plane move? What is its top speed? "

- Trajectory similarity can be measured to determine the distance (or similarities) between a trajectory and a few locations or the difference between two trajectories, while responding KNN queries or grouping trajectories. The length between a current point p and a trajectory A is generally calculated by measuring the length between p and the connected current in A.
- Trajectory data index can be done through augmented R-tree, multi-version R-tree (partition temporal dimension), and grid-based index (partition spatial space).

Uncertainty in a trajectory:

The trajectory data we get is generally a sampling of the object's real movement because the position of a moving vehicle is captured at a specific period. On the other side, an object's motion across two successive sample sites remains uncertain (or referred to be unclear). We want to minimize the ambiguity of a trajectory in this way. On the other side, in such instances, we should have to keep a trajectory much more unclear to prevent a privacy of the user from being disclosed through such trajectories. It is needed to reduce the uncertainty in the trajectory data.

Privacy of trajectory data:

It is more essential to preserve the personal data than to make the trajectory look better [29–31]. As a result, the user's perception is distorted. While guaranteeing the quality of a service, we have to secure a person's trajectory data which leaks private information. Protecting user information from a privacy breach requires us to consider two major situations. One is in real-time endless location-based essential services; in this, an individual may choose not to reveal the exact location while using a provider's service. The spatial and temporal connection among both consecutive samples in a trajectory, in contrast to the straightforward privacy preservation, may help determine the exact location of a user. As a result of this, methodologies such as spatial cloaking [32], mix zones [33], path confusion [34], Euler histogram based on short IDs [35], and dummy trajectory [36] have been designed to protect against information leakage. Second one is biblical trajectories that are published in which attackers may be able to identify a person's home and workplace based on many trajectories. For example, clustering-based methods [37], generalization-based methodologies [38], suppression-based strategies [39], or grid-based techniques are some of the important strategies for helping to protect users' confidentiality in such scenarios. The work [40] provides a detailed study on trajectory confidentiality.

Transfer trajectory to other representations:

- **The transition from trajectory to graph:**
  Trajectories, in addition to it being handled in their initial form, can be converted to various data models. This expands the techniques available for knowledge extraction through trajectories. One of the most common forms of conversion is the conversion of trajectories to graphs. The major effort in performing such a conversion is defining what kind of a location and an edge is in the modified graph. The techniques for converting trajectories into graphs change based on whether to transport or not a network is incorporated in the conversion.

- **The transition from trajectory to matrix:**
  A matrix seems to be another format into which we may convert trajectories. A matrix can be used to supplement incomplete data using current approaches such as CF and MF. A matrix could also be used as an input to detect irregularities. The secret to the transition from trajectory to matrix is found in three areas: (1) What would a row represent? (2) What might a column denote? (3) What would be an entry?

- **The transition from trajectory to tensor:**
  Conversion of trajectories into a (3D) tensor, in which the third dimension is introduced to a matrix to incorporate more knowledge, is a natural continuation of something like the matrix-based conversion. The transformation's objective is generally to replace the missing data or to discover the relationship between two items, such as two road sections or petrol stations. Decomposing a tensor further into combination of a few (low rank) matrices and a central tensor depending on the tensor's nonzero elements is a typical technique to resolve the problem of transition from trajectory to tensor. Whenever a tensor is highly sparse, it is generally broken down with other matrices to get improved results.

3.3.2 Trajectory data mining

When it comes to extracting useful information from large data sources, data mining is a critical element in the development referred to as knowledge discovery [41, 42]. As a general data mining domain, many methods and applications have been explored and analyzed. For example, the work [41] provides a survey of data mining methods for relational and transactional data. There are a variety of mining tasks that can be performed on geographic information [43–45], in addition to the traditional relational and transactional data. The activities that cover are broadly equivalent to those in the generic data mining field, but really the mining techniques have been modified to account for the unique characteristics of geographic information, such as geographical dependencies and the mixture of spatial and non-spatial features. Even though only the activities in any of these investigations are like those in the moving database instance, the review ignores the spatial ordering that is fundamental in location data.

Andrienko *et al.* [46] established a generalized classification system for movement analysis tools. A movement's analysis procedure, according to their paradigm, consists of one or more activities. The many sorts of assignments are classified according to the type of information required, which is specified in terms of the elements concerned, such as things, geography, or temporal, as well as the clear target, that is either a feature or an association. They also categorize assignments according to the level of understanding, which is either fundamental or comprehensive. Castro *et al.* [47] examined the studies on extracting vehicle traces from the standpoint of implementations.

Social dynamics, transportation dynamics, and functional dynamics were used to categorize the examined work. Despite the fact that their assessment included an exhaustive literature evaluation, it only looked at research on taxi traces. As a result, we require a study that includes all implementations of trajectory data mining, rather than just those involving certain types of trajectories. Our research differs with that of [22], in that it considers traces in general and discusses application difficulties which are of great interest in various domains.

Next, take a look at the progress that has been made mostly on application side using the architecture described in Figure 3.5. There are various types of moving vehicles which can be monitored, as shown in the architecture. The monitoring of all these items starts with a set of trajectories and queries, the answers to which could be employed in a variety of applications. Those issues are referred to be application issues since the answers are not applications in and of itself, but information that is applied in application domains.

Various trajectory mining approaches are used to address the application difficulties. If traveling individuals are monitored, for example, we receive their trajectories and certain issues arise, such as the features of such individuals or the geographic location in which they move. We may be able

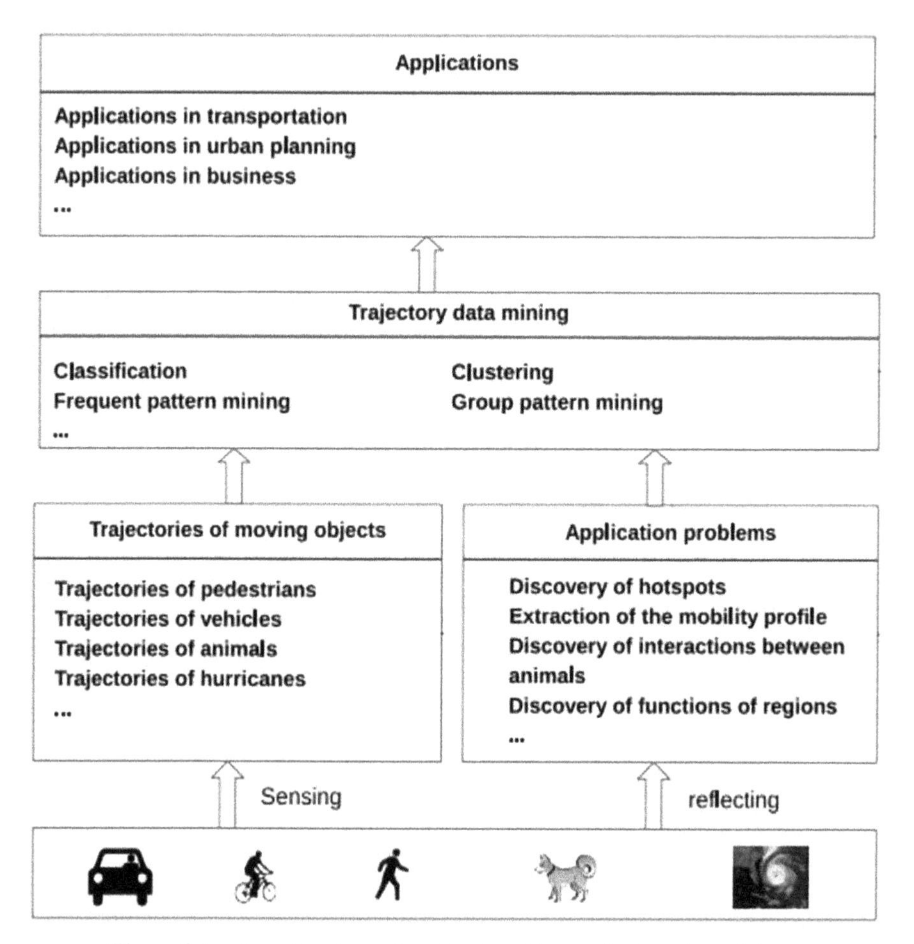

Figure 3.5 Overall outline of application-driven trajectory data mining.

to find areas where numerous people stop often and remain for an extended period by analyzing such trajectories. These access points are defined as locations where people congregate and understanding about hotspots may be useful for a variety of purposes. The identification of hotspots is an applied issue whose answer may be used in a variety of ways like urban computing using trajectory data.

Trajectory mining methods:

Trajectory data mining, like other types of data extraction, attempts to extract useful insight (e.g., patterns) from information that had two major objectives: predictions and descriptions. Prediction is the process of using information to determine uncertain or upcoming quantities of many other variables of interest, whereas description is the process of identifying human-interpretable structures that describe the data. We do not aim to cover all of the available mining methods; instead, we look at the most common ones and see how they relate to one another. By general mining techniques, we imply approaches that can be used on a variety of trajectory data and in a variety of applications. The approaches described are classified into two parts: primary and secondary techniques. Primary approaches seek to categorize trajectories based on characteristics provided by a trajectory data. Secondary mining techniques, from the other side, are hybrid approaches in that they employ a series of primary analysis techniques, traditional statistical techniques, or a mix of the two. As the techniques are discussed, the attention is on the method's aim as well as its basic operating principles as well as the data refinement scale with which it operates (full paths as opposed to partial paths; one path as opposed to many paths; and single path as opposed to many paths). Finally, the debate attempts to identify connections between various techniques.

3.3.2.1 Primary mining methods

To categorize the trajectories based on attributes, primary trajectory extraction techniques are available. Primary mining techniques would be used to produce data for a more in-depth study using other techniques, or they can be used to provide data for a more in-depth examination using a secondary mining approach. There are two primary ways to mine trajectories: trajectory grouping and classification.

Clustering:

In trajectory clustering techniques, trajectories are classified together into fixed number of groups or clusters, according to their movement characteristics. Each cluster's trajectory has direct behaviors that are both different from many other groups' trajectory. For example, statistical and probabilistic frameworks were adjusted to account for the peculiarities of trajectories. By introducing Gaussian noise to mixture models [48], cluster trajectories are prone to be affected by a statutory requirements trajectory. They also utilize hidden Markov models (HMMs) that are best fitted to the trajectory patterns to model a clustering in the same vein as Alon *et al.* [49]. Though trajectory-specific clustering techniques are available, state-of-the-art clustering methods for trajectories are modifications of standard clustering techniques with suitable similarity (or proximity) function definitions. The work [50] contains an overview of conventional clustering techniques.

Depending on the objective of the study, a likeness (or distance) metric such as similar path, similar target, similar origin, similar path and location, or similar directions is used to identify whether trajectories belong to the same cluster. Rokach [51] provides a detailed description of how proximity and likeness functions are used to identify cluster participation. Density-based spatial clustering of applications with noise which is abbreviated as DBSCAN [52] and ordering points to identify the clustering structure which is abbreviated as OPTICS [53] are possibly the two best methods that have already been improved.

A few instances of such expansions are OPTICS [54], which defines a spatio-temporal proximity for comparison and grouping trajectory, and DBSCAN [55], which utilizes two variables for measurement of similarity to optimize the detection of groups and noise. In addition to these examples, it is possible to classify clustering methods in a variety of ways (see, for example, [51, 56, 57]) based on various parameters, although it is hard to point out the differences among them. For most part, categorizing methods include techniques that use division, hierarchy, and density. As a result of partition techniques, all items are grouped into a predetermined number of groups. Objects may be moved between one group to another using such methods, which begin with an arbitrary partition and are improved via repetitions.

One illustration of these kinds of methods is *K*-means [58]. Techniques with a hierarchical organization arrangement of items into clusters and sub-clusters. Clusters are found at a maximum standard when proximity criteria are flexible, while sub-clusters are found when proximity restrictions are tighter. BIRCH (balanced iterative reduction and clustering using hierarchies) [59] is an instance with such a technique. Depending on the density of the items, density-based techniques categorize the items into groups. Beginning with a single object, the cluster expands when more objects appear in the region. For example, DBSCAN falls under this group. If you have several trajectories, you may use trajectory clustering. The feed to the clustering procedure can, therefore, be a single trajectory under certain situations. A particular trajectory is clustered in these situations to characterize a few locations on it [60], in which the locations on the feed trajectory are grouped to find stops on it. It is possible to apply trajectory clusters to complete trajectories or parts of trajectories dependent on the analytic objective and the matching function used in the clustering process. A proximity function described as having an identical route to the destination, for example, will result in the same paths and can be grouped together because the path taken does not matter. As an example, in [61], the focus is now on the overarching characteristics of additional climates. The grouping is performed on segments of trajectories if the focus is on numerous places traveled by the trajectories, for instance, if similarities are characterized as actually visiting the same sorts of sites. TraClus [62] is a clustering method that uses a similar strategy.

Classification:

To classify anything, you need to come up with a rule. A collection of preset classes and a representative collection of entities previously labeled with the class to which they belong to provide the information regarding class assignment. As a result, the sample is known as a training set. It is like classification in general, except trajectories are classified according to the characteristics of the trajectories. It is possible to classify trajectories based on their transport system, for instance, given a limited collection of trajectories that have been explicitly tagged with the mode of transportation utilized.

A typical two-step method is used by many trajectory classification techniques: first obtaining a collection of discriminant features and then utilizing those features that are used to train an existing system classification method. Every trajectory's attribute must be identified to determine which category it resides in. The quantities of these characteristics are, therefore, expressed in a specified way. It is possible for using attributes, namely the trajectories' average speed, average acceleration, period, and size, for instance. When it comes to the discriminative strength of a characteristic, it relies on the sorts of classes involved. When comparing vehicle types, we may say that acceleration has considerably higher detection accuracy than trajectory length.

A simple classification methodology was selected and applied to the retrieved characteristics in the second phase. In [50], the classic classification methods are discussed in depth. To classify trajectories, the following approaches are widely used. Zheng *et al.* [63] utilized the decision tree technique to categorize trajectories into distinct types of transportation. A decision-tree-based learning technique is used to classify the trajectories initially. To address the similar problem of transportation mode categorization, Bolbol *et al.* [64] employed support vector machines (SVMs).

Four characteristics (speed, acceleration, length, and rate of increase in heading) are compared to six forms of transportation in a statistical analysis (bus, car, cycle, train, underground, and walk). To categorize trajectories, they studied equal-sized examples of multiple segments of trajectory using the SVM method. After additional analytic procedures such as separation and grouping have been completed, trajectory classification is achieved. Using trajectory separation and grouping, the TraClass framework [65] extracts region and sub-trajectory characteristics that are subsequently utilized toward support vector machine (SVM) related categorization of the trajectories.

3.3.2.2 Secondary mining methods

However, whereas first-generation mining methods categorize the trajectories, second-generation mining methods focus on the geographical or spatio-temporal organization of the individual trajectories, inside or between groups of trajectories. However, in this section, designers will focus on a few generic methodologies like pattern extraction, outlier identification, and prediction.

Pattern mining:

As the name implies, trajectory pattern recognition seeks to uncover and describe movement patterns that are buried within trajectory data. When and where the pattern happens as well as the entities concerned are described. It has been stated that there are several sorts of motion patterns in the research. The survey of Dodge *et al.* [66] provides a comprehensive perspective of them. These characteristics can also be mined using various other approaches that have been developed. As a result of this, we may classify the approaches into three main categories: repetitive, recurrent, and grouped pattern mining.

A single mobile object's trajectory is used for repetitive pattern mining, whereas many moving objects' paths are used for often occurring pattern mining. The particles do not have to travel at the exact same time and must just visit the same areas roughly in the same order. If you are interested in finding out more about cluster pattern mining, please refer [66]. For example, the daily commute of a commuter or the seasonal migration of birds is an example of repetitive patterns that may be mined. A pattern that repeats itself is also known as a periodic pattern, because the item continues the same path after a certain amount of time has passed. It is difficult to detect periodic patterns since they are imprecise in terms of space and time. For example, an item does not always visit the same position at the same period of the cycle, and the period does not seem to have the same value.

When trying to uncover periodic patterns of movement, it is customary to perform mining to a series of places. Previous research by Cao *et al.* [67] required that a certain time frame was entered into the computer, and then grouped places that were offset from each other by that same time frame to find common areas. Iteratively, they merged the frequent locations they had identified to obtain entire periodic patterns. Although Cao *et al.* [67] expanded prior research [68], both analyses relied on the time as a user input. Using the periodical algorithm [69], this unwanted necessity to define the period beforehand might be overcome by using a more flexible method.

A method combining Fourier transformation and autocorrelation is used to identify the duration automatically in this procedure. As a result, the system discovers citation spots, which are dense locations with more trajectory points than the others and identifies the distinct periods in each. To find periodic trends in movement sequences across reference locations of the same duration, a hierarchical clustering technique with a probabilistic model for distance measurement is employed. Identifying (parts of) paths that are being regularly pursued by moving objects in the trajectory database is called frequent pattern analysis. The geographical or spatio-temporal features of the trajectories can be used to identify frequent trajectory patterns [70]. Using spatial features as a criterion, the sequence of sites visited is the sole factor considered.

Spatio-temporal sequential patterns [71] and GSPs [72] are examples. Using spatio-temporal features, the order in which sites are visited is considered, as is the transition time among locations. Finding significant sections from the trajectory data and then performing sequential mining to a chronologically indexed series of these parts is one popular method for mining frequent patterns. Giannotti *et al.* [73] proposed two methods for extracting T-patterns in this manner. A discretization of the space is the first approach when it comes time to mine the sequences [74]. Utilizing trajectory sections, the second technique interactively identifies objects of interest that are then translated into sequences of areas out of which recurrent T-patterns are constructed. When using group pattern extraction, you are trying to find patterns in the movement of a collection of elements that are moving simultaneously, the items in a group must stay near to one other in space for a long length of time. It has been determined that a number of group patterns and its varieties may be derived from the general assessment of spatial and temporal proximity, and its internal workings and individual feature flocks [75], convoys [76], and swarms [77] have been the most investigated patterns.

A structure wherein the individuals of a group communicate and every individual has a particular role falls under the category of group patterns. The leadership pattern [78] is an instance of this kind of structure. In the pursuing pattern [79], the item going forward seeks to escape the follower, who is trying to catch up to it, to get away. The work [80] provides a variety of instances of similar practice under the broad idea of relative motion (REMO). Clustering techniques and testing the constraint on factors that characterize the design, like the minimal number of group participants and the minimal length of the structure, are common ways to group pattern mining. When it comes to mine flocks and convoys, for example, density-based clustering has been used on them in [81]; when it comes to mine swarms in [82]. An initial time-stamp-based clustering approach is used to identify geographical clusters that share a percentage of objects, irrespective of the temporal distance.

Outlier detection:

It is the goal of outlier identification in trajectory analysis to discover trajectories that deviate from the overall trend of the trajectory collection. For example, outlier identification focuses on uncommon paths (i.e., not matching the common path pursued by most other trajectories), whereas pattern recognition concentrates on frequent patterns in the trajectory collection. Tracking outliers in trajectories is analogous to trajectory classification in that it may be done by analyzing either complete trajectories or portions of them. On the other hand, [83, 84] provides instances of outlier identification applicable to sub-trajectories. A proximity metric or clustering approach is being used to find outliers in sub-trajectories. A last step identifies which sub-trajectories corresponding to which outlying sub-trajectories [83] have used this method.

Other mining processes, particularly grouping, might produce outliers. Many outlier identification methods in the literature employ some sort of clustering technique to identify items that are not present in any clusters as outliers. Determining the right proximity metric to distinguish between these divergent paths is therefore crucial. The neighborhood within each trajectory can be analyzed by tracking the number of neighbors or by applying a density-based clustering algorithm to find outliers. A trajectory that has very few neighbors is classified as an outlier. IBAT [85] is an expanded version of this technique. IBAT partitions the research region together into grid structure and groups the trajectories that pass the very same source–destination cell combination (source/destination cell pairs). As a final step, the technique relies on an isolation mechanism to detect outlier trajectories, which makes use of the fact that they are rare and distinct within their group. Pathway outlier identification may also be done using classification techniques. One method is to collect from the trajectory a set of specified characteristics, after which the derived vectors are subjected to a distance measure. Distance measurements were used to four characteristics (direction, speed, slant, and position) in [84] to find outliers in the trajectory. Build a two-label classifier model with one label for usual trajectories and the other for aberrant trajectories.

Prediction:

To determine the correct location of an object in motion, existing trajectory data is used. As a result of the fast-growing. A prominent use of this technology is location-based services [50]. The main goal of this study on trajectory-based prediction aims to forecast a place (destination). However, there really are studies that attempt to forecast the complete path, based on a road system. When predicting a place based on the trajectory of an object trajectories. The research on location prediction is characterized by two strategies (Markov models as well as sequential rules or trajectory patterns) and it can be divided into three categories:

i. Those that consider just the information of the involved moving item [86, 87].

ii. Those that consider just the data of some of the other moving objects [88].

iii. Composite approaches that consider both information of the concerned item and the information of many other moving objects. For location prediction, studies based on Markov models employ probabilistic models. When predicting routes as opposed to locations, a sequence of route segments must be produced to arrive at the destination.

The three most prevalent techniques to route prediction are as follows.

- Trip observations: A driver observes a series of excursions. To predict a path for a new trip, this method considers the start location and maybe a short beginning segment of the new journey and then attempts to identify a suitable connection with another trip [89].
- Markov model: The driver's long-term journey histories are used to develop a probability-based model for forecasting the driver's short-term route. Using the driver's just-driven path as a guide, the algorithm forecasts the upcoming road section.
- Turn proportions: As a result of this method, the route is constructed by forecasting whichever way a driver would go at each crossing in turn. When Krumm [90] makes a turn prediction, he considers the fraction of cars that would make each choice at a crossing, as well as the driving times among pairs of road segments. On the hypothesis that drivers prefer to choose another route with the most destinations available and that they take the fastest path to their target, this technique assumes that drivers should choose the route with the most choices.

Here are the correlations between many of the mining methods we have examined so far, and what they mean for you. In Table 3.1, the primary methods and subsidiary methods are presented as headings for columns and rows, accordingly. If there is an intersection between two mining methods, then the secondary mining technique on top of a column will call on the primary mining method to do what is needed. To find outlier trajectories, for example, the outlier detection approach may use grouping to group trajectories or

Table 3.1 Associations seen between approaches of trajectory analysis.

		Secondary methods		
		Pattern mining	**Outlier detection**	**Mining prediction**
Primary methods	**Clustering**	• [67, 91] Grouping locations that occur at the same time. • It is possible to group similar trajectories [92, 93]. • Extraction of locations of interest in order to perform pattern mining [73]. • In order to find sequential patterns, close places can be aggregated [94].	• Clustering trajectories or sub-trajectories that are similar [80].	• Clustering users who are identical [95]. • It is useful to categorize comparable travels of a user [96]. • In order to create a trajectory pattern, users can cluster their stopover spots [97]. • Creating periodic structures by grouping visiting areas [87];
	Classification		• Normal and abnormal trajectory characterization [99].	• Classifying an ongoing route into predetermined [97] one-trajectory groups

sub-trajectories that are comparable. Furthermore, the referencing entry [83] shows an explanation of this association. There is no pattern mining categorization in the research since there is no empty table cell.

3.4 Applications of Trajectories

Trajectory data analysis generates information that can be used in a variety of fields. Application problems can be solved in a variety of ways. A person's movement profile can also be used in urban planning, for example, to manage recreational opportunities [100], in businesses to plan promotional strategies, or even in mobility to set up a car communication platform.

Transportation:
When working in the transportation industry, it is almost always essential to recognize the features of areas and regions and traveling objects themselves. They relate to the application concerns of characterizing locations (mostly hotspot identification) and characterizing moving objects (primarily assessing people's mobility profiles). Issues with trajectory-based advice and prediction can also occur in specific instances. There are a number of applications of trajectory data analysis in transportation, such as optimizing the experience of driving, cab and transit service, organizing and executing transportation infrastructure, and travel strategic planning [9].

Urban planning:
According to city designers, in the analysis to define the prevalence and features of a relationship among urban zones, they must conduct employment surveys. When it comes to characterizing regions, it is all about mining trajectories to recognize spatial activities or field use (e.g., [101]). Another concern is detecting and describing regional connectedness. Trajectory data mining has been used in the past to explore urban borderline evolution and uncover flaws in a transport system [102].

Environment:
Pollution control requires measuring pollution levels at multiple sites and combining environmental data with other data. In the context of environmental pollution or sound pollution [103], these have been converted into the problem of describing locations and regions by mining trajectories that relate to measurement values at various locations. Tracking data analysis is a technique for growing the number of observed locations, combining them, and evaluating them [102].

Energy:
When it comes to the power generation, it is important to know how particular areas and populations use energy. The extraction trajectories of automobiles [104] resolve these challenges by characterizing regions and moving objects. Developing eco-driving feedback channels and selecting sites for electric car charging infrastructure have both used this method.

Online social networks:
Using trajectory data mining, you may learn about people's behavior and compare it with other people's behavior to come up with better recommendations. Problems related to detecting individuals' movement profiles, finding and describing their social contacts, and ultimately trajectory-based suggestion are discussed. Examples include recommending friends [105] based on similarities derived from movement profiles or geography and route. It is also possible to analyze personal and collective lifestyles by looking at the actions of individuals based on their trajectories.

Business:
In the corporate environment, it is almost always necessary to estimate the visit possibilities of specific locations and the people's choices of movement patterns in these locations. Using trajectory data mining, the implementation difficulties of identifying places or areas as well as moving vehicles are addressed. For example, best business locations [66, 125], best advertising sites [67], and ideal store layouts [52, 72] have been advised utilizing trajectory data mining.

Public safety:
Apps in this category must be able to recognize and detect places or moving objects which have the capacity to create or be affected by challenges to public safety and security. Data analytics can help identify or even foresee circumstances that threaten security and stability because they are generally associated with large movements. Classifying places and moving vehicles, which can transcend to trajectory-based forecasting, are typical application challenges that must be solved.

Ecology:
Individual animals' behavior, their interactions, and the habitat's utilization are primary concerns for those who manage or conduct animal-related research. When it comes down to it, our study solves an important problem in computer science: How to characterize individual moving objects while

detecting and describing their relationships? It has enabled automatic identification of animal activity as well as their communication [106] and comprehending their habitat utilization [30], for example.

Sports analytics is a growing field:
With so much motion and action, sports can be difficult to evaluate in real time. The main objective is to extract and evaluate a person's dynamic characteristics and his relationships with some other competitors to make informed decisions. As a result, moving objects and its interconnections can be characterized. The mobility and relationships [107] of individuals have been analyzed using individual trajectories.

3.5 Issues for Trajectory Data Mining

Trajectory data mining has caught the interest of researchers, and considering potential uses, it is certain to do so in the future. Furthermore, trajectory data mining is mostly growing in breadth, with most of the potential effort focusing on producing variations. Utilizing methodological approaches and extending technique variations to similar application issues, there are many unresolved questions that must be addressed to capture trajectory data mining toward the next stage.

Problems that must be resolved:
Several of the concerns are addressed in this section.

- **Dealing with enormous amounts of data:**
  Most of the mining techniques used on trajectories have been created in a time when information was scarce and/or in a particular level. Because trajectory data is intrinsically enormous and has become more easily available, research development on trajectory data analysis ought to be adaptable to manage enormous amounts of data. This necessitates, among several other things, highly appropriate crawling and aggregated approaches that reduce data while avoiding risk of data loss.

- **Data integration and diversity management:**
  Many research findings on trajectory data mining are focused on a complete type of dataset (e.g., taxi GPS-based tracks). Nevertheless, as recognized in several articles [101], these investigations confront the issue of incomplete data in particular locations or during certain seasons, which might be addressed by combining other types of information

(e.g., GSM-based trajectory, location specific media, and so on). As a result, future research should focus on incorporating multiple forms of data while resolving problems posed by their diversity, such as harmonizing disparate data types and sample rates.

- **Exploration of cross-scale trajectory data:**
  A trajectory data mining approach is typically designed as well as proven on a single dataset at a single size and precision. Furthermore, as proven in [108], the size and precision with which the approach is performed can have an impact on the outcome. As a result, future research on trajectory data mining technique must include size and precision to assess their influence on process parameters.
- **Mining technique experts and implementation domain experts:**
  While investigators in trajectory data mining algorithms have primarily focused on that spatial trajectory with really no regard for application environment, latest collaborative project attempts have suggested that the best outcomes with real-world relevance of trajectory data analysis can be obtained through cooperation between technique experts and domain knowledge specialists. This type of partnership makes it easier for methodologists to have access to important data. This can assist technique specialists in determining appropriate choices for variables used in mining techniques, a process that is typically left to the technique user without help.
- **Handling of personal information:**
  Although animal trajectories are simply publicly revealed, human monitoring relates to severe confidentiality, rendering highly skilled human trajectory data difficult to get for processing technique developers. Although there are efforts to resolve such difficulties (for example, anonymization [109]), they persist. As a result, future research on trajectory data mining must focus on developing mining approaches which are more reasonable in terms of managing the collection of specific knowledge with the privacy of monitored persons.

3.6 Publicly Available Trajectory Datasets

- GeoLife Trajectory Dataset (GeoLife Data): A GPS trajectory dataset gathered by 182 individuals from April 2007 to August 2012 from the Microsoft Research GeoLife project. The database is being used to assess user proximity allowing for contact and region suggestions. This is also utilized to investigate the challenge of determining the closest trajectory to a series of querying locations.

- T-Drive Taxi Trajectories (T-Drive Data): A selection of trajectories from the Microsoft Research T-Drive research created by more than 10,000 taxicabs in Beijing during a week in 2008. The entire original data has been used to propose the essentially quickest travel information to regular drivers and suggest passenger-pickup locations for cab drivers.
- GPS Trajectory using Transportation Labeling (Trajectory using Means Of transportation): Every trajectory includes travel mode indicators which include driving, riding the bus, riding bikes, and having to walk. The database is suitable for evaluating trajectory categorization and movement recognition.
- Check-in Information from Location-Based Social Networking Sites (Customer Check-in Information): The data contains check-in data from over 49,000 individuals in New York City and 31,000 individuals in Los Angeles, and the individuals' social structures. Each check-in comprises a venue ID, venue type, current time, and unique identifier. Because a participant's check-in information may be viewed as a low-sampling-rate trajectory, such information has really been utilized to investigate trajectory ambiguity and assess position suggestion.
- Hurricane Tracking [Hurricane Tracking (HURDAT)]: The National Hurricane Service (NHS) provides this database, which contains 1740 trajectories of Atlantic hurricanes (officially designated as tropical cyclones) from 1851 to 2012. In addition, NHS includes descriptions on normal storm paths for every month of the yearly storm season, which runs from June to November. The information would be used to experiment with trajectory grouping and uncertainties.
- The Greek Truck Trajectories (The Greek Trucks Database): This database contains 1100 trajectories from 50 distinct trucks carrying mortar across Athens, Greece. This is utilized to assess the trajectory pattern matching challenge.
- Movebank Animal Monitoring Data (Movebank Data Set): Movebank is a public, online resource of animal tracking data that assists scientists in managing, sharing, protecting, analyzing, and archiving their information.

3.7 Conclusion and Future Work

The vast amount of data created in urban areas, combined with improvements in computer technologies, has given us with unparalleled opportunity to address the major issues that cities confront. Urban computing seems to be an interdisciplinary field in which technology disciplines are combined with traditional city-related fields including such civil engineering, ecology, psychology, economics, and energy. In the framework of towns, the goal of urban computing is to capture, incorporate, and analyze large amounts of trajectory data to enhance urban areas and living quality which will result in smarter, cleaner cities that will benefit millions and millions of people. As a result of trajectory data, there will be a blurring of the borders among traditional computer science areas (such as databases, algorithms, and visualization) or perhaps a bridge across various fields (e.g., computer sciences and civil engineering). It is true that urban computing does have the potential to change urban technologies and advancement. However, relatively a few approaches have yet to be investigated, including hybrid indexing structures for transmission data and information convergence throughout heterogeneous data sources.

This article examines the strategies used in different phases of trajectory data mining in urban computing, categorizing them, and comparing them. This article also proposes methods for converting raw trajectory data into useful information and the application of current data mining methods to new data structures. An overview is provided on how to extract the potential of knowledge from trajectories in this article. Researchers and experts from a variety of fields, including computer sciences, a wider spectrum of communities dealing with trajectory-related problems, can make use of this article for further developments. This essay will conclude with a public dataset being listed, and some future directions being suggested. In addition to that, in future, it is necessary to improve the methods for smart transportation, detecting urban anomalies, the intelligent environment, and so on. Finally, this article explained the idea, structure, and problems of urban computing that offered sample implementations and methodologies for urban computing and recommended a few research paths that require community involvement.

References

[1] Spinsanti, L., Berlingerio, M., and Pappalardo, L. Mobility and geo-social networks. In Mobility Data, C. Renso, S. Spaccapietra, and E. Zimanyi, Eds. Cambridge University Press, 2013, pp. 315–333. doi:10.1017/cbo9781139128926.017.

[2] Pelekis, N., and Theodoridis, Y. Mobility Data Management and Exploration. Springer-Verlag New York, 2014. doi:10.1007/978-1-4939-0392-4.

[3] Phithakkitnukoon, S., Veloso, M., Bento, C., Biderman, A., and Ratti, C. Taxi-aware map: Identifying and predicting vacant taxis in the city. In Ambient Intelligence, LNCS 6439, B. de Ruyter, R. Wichert, D. V. Keyson, P. Markopoulos, N. Streitz, M. Divitini, N. Georgantas, and A. M. Gomez, Eds. Springer, 2010, pp. 86–95. doi:10.1007/978-3-642-16917-5_9.

[4] J. Bao, Y. Zheng, and M. F. Mokbel. 2012. Location-based and preference-aware recommendation using sparsegeo-social networking data. In *Proceedings of the 20th ACM SIGSPATIAL International Conference onAdvances in Geographic Information Systems*. ACM, 199–208.

[5] J. Bao, Y. Zheng, D. Wilkie, and M. F. Mokbel. 2015. A survey on recommendations in location-based social networks. *GeoInformatica*, 19, 3, 525–565.

[6] L. X. Pang, S. Chawla, W. Liu, and Y. Zheng. 2011. On mining anomalous patterns in road traffic streams. In *Proceedings of the International Conference on Advanced Data Mining and Applications*. 237–251.

[7] W. Liu, Y. Zheng, S. Chawla, J. Yuan, and X. Xie. 2011. Discovering spatio-temporal causal interactions in traffic data streams. In *Proceedings of the 17th ACM SIGKDD International Conference on Knowledge Discovery and Data Mining*. ACM, 1010–1018.

[8] L. A. Tang, Y. Zheng, X. Xie, J. Yuan, X. Yu, and J. Han. 2011. Retrieving k-nearest neighboring trajectories by a set of point locations. In *Proceedings of the 12th Symposium on Spatial and Temporal Databases*.Springer, 223–241.

[9] Y.Wang, Y. Zheng, and Y. Xue. 2014. Travel time estimation of a path using sparse trajectories. In *Proceedings of the 20th ACM SIGKDD International Conference on Knowledge Discovery and Data Mining*. ACM,25–34.

[10] J. Yuan, Y. Zheng, X. Xie, and G. Sun. 2013a. T-Drive: Enhancing driving directions with taxi drivers'intelligence. *IEEE Transaction on Knowledge and Data Engineering* 25, 1 (2013), 220–232.

[11] N. J. Yuan, Y. Zheng, L. Zhang, and X. Xie. 2013b. T-Finder: A recommender system for finding passengersand vacant taxis. *IEEE Transaction on Knowledge and Data Engineering* 25, 10 (2013), 2390–2403.

[12] J. Yuan, Y. Zheng, X. Xie, and G. Sun. 2011a. Driving with knowledge from the physical world. In *Proceedings of the 17th ACM SIGKDD International Conference on Knowledge Discovery and Data Mining*. ACM,316–324.

[13] J. G. Lee, J. Han, and K. Y. Whang. 2007. Trajectory clustering: A partition-and-group framework. In *Proceedings of the ACM SIGMOD Conference on Management of Data.* ACM, 593–604.
[14] J. Yuan, Y. Zheng, X. Xie, and G. Sun. 2011b. Driving with knowledge from the physical world. In Proceedings of 17th SIGKDD Conference on Knowledge Discovery and Data Mining. ACM, 316–324.
[15] J. Yuan, Y. Zheng, X. Xie, and G. Sun. 2013b. T-Drive: Enhancing driving directions with taxi drivers' intelligence. Transactions on Knowledge and Data Engineering 25, 1, 220–232.
[16] Y. Zheng and O. Wolfson. 2012c. Proceedings of the 1st International Workshop on Urban Computing. In conjunction with KDD 2012.
[17] B. Pan, Y. Zheng, D. Wilkie, and C. Shahabi. 2013. Crowd sensing of traffic anomalies based on human mobility and social media. In Proceedings of the 21th ACM SIGSPATIAL Conference on Advances in Geographical Information Systems. ACM.
[18] J. Yuan, Y. Zheng, L. Zhang, X. Xie, and G. Sun. 2011a. Where to find my next passenger? In Proceedings of 13th ACM International Conference on Ubiquitous Computing. ACM, 109–118.
[19] Y. Zheng. 2011a. Location-based social networks: Users. In Computing with Spatial Trajectories, Y. Zheng and X. Zhou, Eds. Springer, 243–276.
[20] Y. Zheng, L. Zhang, Z. Ma, X. Xie, and W.-Y. Ma. 2011c. Recommending friends and locations based on individual location history. ACM Transaction on the Web 5, 1 (2011), 5–44.
[21] Y. Zheng and X. Xie. 2011b. Learning travel recommendations from user-generated GPS traces. ACM Transactions on Intelligent Systems and Technology 2, 1 (2011), 2–19.
[22] Y. Ye, Y. Zheng, Y. Chen, J. Feng, and X. Xie. 2009. Mining individual life pattern based on location history. In Proceedings of the 10th IEEE International Conference on Mobile Data Management. IEEE, 1–10.
[23] N. J. Yuan, Y. Zheng, L. Zhang, and X. Xie. 2013b. T-Finder: A recommender system for finding passengers and vacant taxis. IEEE Transaction on Knowledge and Data Engineering 25, 10 (2013), 2390–2403.
[24] W.-C. Lee and J. Krumm. 2011. Trajectory pre-processing. Computing with Spatial Trajectories, Y. Zheng and X. Zhou (Eds.). Springer, 1–31.
[25] J. G. Lee, J. Han, and K. Y. Whang. 2007. Trajectory clustering: A partition-and-group framework. In Proceedings of the ACM SIGMOD Conference on Management of Data. ACM, 593–604.
[26] J. Lee, J. Han, and X. Li. 2008. Trajectory outlier detection: A partition-and-detect framework. In Proceedings of the 24th IEEE International Conference on Data Engineering. IEEE, 140–149.

[27] J. Krumm. 2011. Trajectory analysis for driving. Computing with Spatial Trajectories, Y. Zheng and X. Zhou (Eds.). Springer, 213–241.
[28] K. Deng, K. Xie, K. Zheng, and X. Zhou. 2011. Trajectory indexing and retrieval. *Computing with Spatial Trajectories*. Y. Zheng and X. Zhou (Eds.). Springer, 35–60.
[29] O. Abul, F. Bonchi, and M. Nanni. 2008. Never walk alone: Uncertainty for anonymity in moving objects databases. In Proceedings of the 24th IEEE International Conference on Data Engineering. IEEE, 376–385.
[30] A. Y. Xue, R. Zhang, Y. Zheng, X. Xie, J. Huang, and Z. Xu. 2013. Destination prediction by sub-trajectory synthesis and privacy protection against such prediction. In Proceedings of the 29th IEEE International Conference on Data Engineering. IEEE, 254–265.
[31] C. Y. Chow and M. F. Mokbel. 2011. Privacy of spatial trajectories. Computing with Spatial Trajectories, Y. Zheng and X. Zhou (Eds.). Springer, 109–141.
[32] M. F. Mokbel, C. Y. Chow, and W. G. Aref. 2007. The new Casper: Query processing for location services without compromising privacy. In Proceedings of the 23rd IEEE International Conference on Data Engineering. IEEE, 1499–1500.
[33] A. R. Beresford and F. Stajano. 2003. Location privacy in pervasive computing. IEEE Pervasive Computing 2, 1 (2003), 46–55.
[34] B. Hoh, M. Gruteser, H. Xiong, and A. Alrabady. 2010. Achieving guaranteed anonymity in GPS traces via uncertainty-aware path cloaking. IEEE Transactions on Mobile Computing 9, 8 (2010), 1089–1107.
[35] H. Xie, L. Kulik, and E. Tanin. 2010. Privacy-aware traffic monitoring. IEEE Transactions on Intelligent Transportation Systems 11, 1 (2010), 61–70.
[36] H. Kido, Y. Yanagisawa, and T. Satoh. 2005. An anonymous communication technique using dummies for location-based services. In Proceedings of the 3rd International Conference on Pervasive Services. IEEE, 88–97.
[37] . Abul, F. Bonchi, and M. Nanni. 2008. Never walk alone: Uncertainty for anonymity in moving objects databases. In Proceedings of the 24th IEEE International Conference on Data Engineering. IEEE, 376– 385.
[38] M. E. Nergiz, M. Atzori, Y. Saygin, and B. Guc. 2009. Towards trajectory anonymization: A generalization-based approach. Transactions on Data Privacy 2, 1 (2009), 47–7
[39] M. Terrovitis and N. Mamoulis. 2008. Privacy preservation in the publication of trajectories. In Proceedings the 9th IEEE International Conference on Mobile Data Management. IEEE, 65–72.

[40] C. Y. Chow and M. F. Mokbel. 2011. Privacy of spatial trajectories. Computing with Spatial Trajectories, Y. Zheng and X. Zhou (Eds.). Springer, 109–141.

[41] Fayyad, U., Piatetsky-Shapiro, G., and Smyth, P. From data mining to knowledge discovery: an overview. In Advances in Knowledge Discovery and Data Mining, U. Fayyad, G. Piatetsky-Shapiro, Amith, P. Smyth, and R. Uthurusamy, Eds. AAAAI Press, 1996, pp. 1–34.

[42] Maimon, O., and Rokach, L. Introduction to knowledge discovery and data mining. In Data Mining and Knowledge Discovery Handbook, O. Maimon and L. Rokach, Eds., 2nd ed. Springer, 2010, pp. 1–15. doi:10.1007/978-0-387-09823-4_1.

[43] Mennis, J., and Guo, D. Spatial data mining and geographic knowledge discovery-an introduction. Computers, Environment and Urban Systems 33, 6 (2009), 403–408. Doi: 10.1016/j.compenvurbsys.2009.11.001.

[44] Miller, H. J. Geographic data mining and knowledge discovery. In Handbook of Geographic Information Science, J. P. Wilson and A. S. Fotheringham, Eds. Blackwell Publishing, 2008, pp. 352–366. doi:10.1002/9780470690819.ch19.

[45] Miller, H. J., and Han, J. Geographic data mining and knowledge discovery: An overview. In Geographic Data Mining and Knowledge Discovery, H. J. Miller and J. Han, Eds., 2nd ed. Taylor and Francis Group, FL, USA, 2009, pp. 1–26. doi:10.4324/9780203468029_chapter_1.

[46] Andrienko, G., Andrienko, N., Bak, P., Keim, D., Kisilevich, S., and Wrobel, S. A conceptual framework and taxonomy of techniques for analyzing movement. Journal of Visual Languages and Computing 22, 3 (2011), 213–232. doi:10.1016/j.jvlc.2011.02.003.

[47] Castro, P. S., Zhang, D., Chen, C., Li, S., and Pan, G. From taxi GPS traces to social and community dynamics: A survey. ACM Computing Surveys 46, 2 (2013), 1–34. doi:10.1145/2543581.2543584.

[48] Gaffney, S., and Smyth, P. Trajectory clustering with mixture of regression models. In KDD' 99: Proc. Fifth International Conference on Knowledge Discovery and Data Mining (1999), ACM Press, pp. 63–72. doi:10.1145/312129.312198.

[49] Alon, J., Sclaroff, S., Kollios, G., and Pavlovic, V. Discovering clusters in motion time series data. In Proc. IEEE Computer Society Conference on Computer Vision and Pattern Recognition (2003), IEEE, pp. 375–381. doi:10.1109/cvpr.2003.1211378.

[50] Han, J., Kamber, M., and Pei, J. Data mining: concepts and techniques, 3rd ed. Morgan Kaufmann, 2012. doi:10.5860/choice.49-3305.

[51] Rokach, L. A survey of clustering algorithms. In Data Mining and Knowledge Discovery Handbook, O. Maimon and L. Rokach, Eds., 2nd ed. Springer, 2010, pp. 269–298. doi:10.1007/978-0-387-09823-4_14.
[52] Ester, M., Kriegel, H.-P., Sander, J., and Xu, X. A density-based algorithm for discovering clusters in large spatial databases with noise. In KDD'96: Proc. Second International Conference on Knowledge Discovery and Data Mining (1996), pp. 226–231.
[53] Ankerst, M., Breunig, M. M., Kriegel, H.-P., and Sander, J. Optics: Ordering points to identify the clustering structure. In Proc. ACM SIGMOD International Conference on Management of Data (1999), ACM Press, pp. 49–60. doi:10.1145/304182.304187.
[54] Nanni, M., and Pedreschi, D. Time-focused clustering of trajectories of moving objects. Journal of Intelligent Information Systems 27, 3 (2006), 267–289. doi:10.1007/s10844-006-9953-7.
[55] Birant, D., and Kut, A. St-Dbscan: An algorithm for clustering spatial-temporal data. Data and Knowledge Engineering 60, 1 (2007), 208–221. doi:10.1016/j.datak.2006.01.013.
[56] Fraley, C., and Raftery, A. E. How many clusters? which clustering method? answers via model-based cluster analysis. The Computer Journal 41, 8 (1998), 578–588. doi:10.1093/comjnl/41.8.578.
[57] Han, J., Lee, J.-G., and Kamber, M. An overview of clustering methods in geographic data analysis. In Geographic Data Mining and Knowledge Discovery, H. Miller and J. Han, Eds., 2nd ed. Taylor and Francis Group, 2009, pp. 149–187. doi:10.1201/9781420073980.ch7.
[58] MacQueen, J. Some methods for classification and analysis of multivariate observations. In Proc. 5th Berkley Symposium on Mathematical Statistics and Probability, L. L. Cam and J. Neyman, Eds., vol. 1. University of California Press, 1967, pp. 281–297.
[59] Zhang, T., Ramakrishnan, R., and Livny, M. Birch: An efficient data clustering method for very large databases. In SIGMOD '96: Proc. 1996 ACM SIGMOD International Conference on Management of Data (1996), ACM, pp. 103–114. doi:10.1145/235968.233324.
[60] Gaffney, S. J., Robertson, A. W., Smyth, P., Camargo, S. J., and Ghil, M. Probabilistic clustering of extratropical cyclones using regression mixture models.*Climate Dynamics 29*, 4 (2007), 423–440. doi:10.1007/s00382-007-0235-z.
[61] Palma, A. T., Bogorny, V., Kuijpers, B., and Alvares, L. O. A clustering-based approach for discovering interesting places in trajectories. In SAC '08: Proc. 2008 ACM symposium on Applied computing (2008), ACM, pp. 863–868. doi:10.1145/1363686.1363886.

[62] Lee, J.-G., Han, J., and Whang, K.-Y. Trajectory clustering: A partition-andgroup framework. In *roc. 2007 ACM SIGMOD International Conference on Management of Data* (2007), ACM Press, pp. 593–604. doi:10.1145/1247480.1247546.

[63] Zheng, Y., Chen, Y., Li, Q., Xie, X., and Ma, W.-Y. Understanding transportation modes based on gps data for web applications. *ACM Transactions on the Web 4*, 1 (2010), 1–36. doi:10.1145/1658373.1658374

[64] Bolbol, A., Cheng, T., Tsapakis, I., and Haworth, J. Inferring hybrid transportation modes from sparse GPS data using a moving window SVM classification. *Computers, Environment and Urban Systems 31*, 6 (2012), 526–537. doi:10.1016/j.compenvurbsys.2012.06.001.

[65] Lee, J.-G., Han, J., Li, X., and Gonzalez, H. Traclass: Trajectory classification using hierarchical region based and trajectory-based clustering. *Proceedings of the VLDB Endowment 1*, 1 (2008), 1081–1094. doi:10.14778/1453856.1453972.

[66] Dodge, S., Weibel, R., and Lautenschütz, A.-K. Towards a taxonomy of movement patterns. *Journal of Information Visualization 7* (2008), 240–252. doi:10.1057/palgrave.ivs.9500182.

[67] Cao, H., Mamoulis, N., and Cheung, D. W. Discovery of periodic patterns in spatiotemporal sequences. *IEEE Transactions on Knowledge and Data Engineering 19*, 4 (2007), 453–467. doi:10.1109/tkde.2007.1002.

[68] Mamoulis, N., Cao, H., Kollios, G., Hadjieleftheriou, M., Tao, Y., Andcheung, D. W. Mining, indexing, and querying historical spatiotemporal data. In*KDD '04: Proc. 2004 ACM SIGKDD international conference on Knowledge discovery anddata mining* (2004), ACM, pp. 236–245. doi:10.1145/1014052.1014080.

[69] Li, Z., Ding, B., Han, J., Kays, R., and Nye, P. Mining periodic behaviors for moving objects. In *KDD '10: Proc. 16th ACM SIGKDD international conference on Knowledge discovery and data mining* (2010), ACM Press, pp. 1099–1108. doi:10.1145/1835804.1835942.

[70] Körner, C., May, M., and Wrobel, S. Spatiotemporal modeling and analysis - introduction and overview. *KI - KünstlicheIntelligenz 26*, 3 (2012), 215–221. doi:10.1007/s13218-012-0215-2.

[71] Cao, H., Mamoulis, N., and Cheung, D. W. Mining frequent spatio-temporal sequential patterns. In *ICDM: Proc. Fifth IEEE International Conference on DataMining*(2005), IEEE Computer Society Press, pp. 82–89. doi:10.1109/icdm.2005.95.

[72] Orellana, D., Bregt, A. K., Ligtenberg, A., and Wachowicz, M. Exploring visitor movement patterns in natural recreational areas. *Tourism Management 33*, 3 (2012), 672–682. doi:10.1016/j.tourman.2011.07.010.

[73] Giannotti, F., Nanni, M., Pinelli, F., and Pedreschi, D. Trajectory pattern mining. In *KDD'07: Proc. 13th International Conference on Knowledge Discovery and Data Mining* (2007), ACM Press, pp. 330–339. doi:10.1145/1281192.1281230.

[74] Giannotti, F., Nanni, M., and Pedreschi, D. Efficient mining of temporally annotated sequences. In *Proc. 2006 SIAM International Conference on Data Mining* (2006), Society for Industrial & Applied Mathematics (SIAM), pp. 348–359. doi:10.1137/1.9781611972764.31.

[75] Benkert, M., Gudmundsson, J., Hübner, F., and Wolle, T. Reporting flock patterns. *Computational Geometry 41*, 3 (2008), 111–125. doi:10.1016/j.comgeo.2007.10.003.

[76] Jeung, H., Yiu, M. L., Zhou, X., Jensen, C. S., and Shen, H. T. Discovery of convoys in trajectory databases. *Proceedings of the VLDB Endowment 1*, 1 (2008), 1068– 1080. doi:10.14778/1453856.1453971.

[77] Li, Z., Ding, B., Han, J., and Kays, R. Swarm: mining relaxed temporal moving object clusters. *Proceedings of the VLDB Endowment 3*, 1-2 (2010), 723–734. doi:10.14778/1920841.1920934.

[78] Andersson, M., Gudmundsson, J., Laube, P., and Wolle, T. Reporting leaders and followers among trajectories of moving point objects. *Geoinformatica 12*, 4 (2008), 497–528. doi:10.1007/s10707-007-0037-9.

[79] De Lucca Siqueira, F., and Bogorny, V. Discovering chasing behavior in moving object trajectories. *Transactions in GIS 15*, 5 (2011), 667–688. doi:10.1111/j.1467-9671.2011.01285.x.

[80] Laube, P., Van Kreveld, M., and Imfeld, S. Finding remo – detecting relative motion patterns in geospatial lifelines. In *Developments in Spatial Data Handling*, P. F. Fisher, Ed. Springer, 2005, pp. 201–215. doi:10.1007/3-540-26772-7_16.

[81] Vieira, M. R., Bakalov, P., and Tsotras, V. J. On-line discovery of flock patterns in spatio-temporal data. In *GIS '09: Proc. 17th ACM SIGSPATIAL International Conference on Advances in Geographic Information Systems* (2009), ACM, pp. 286–295. doi:10.1145/1653771.1653812.

[82] Yu, Y., Wang, Q., Kuang, J., and He, J. Tgcr: An efficient algorithm for mining swarm in trajectory databases. In *Proc. 2011 IEEE International Conference on Spatial Data Mining and Geographical Knowledge Services* (2011), IEEE, pp. 90–95. doi:10.1109/icsdm.2011.5969011.

[83] Lee, J.-G., Han, J., and Li, X. Trajectory outlier detection: A partition-and-detect framework. In *ICDE 2008: Proc. International Conference on Data Mining* (2008), IEEE, pp. 140–149. doi:10.1109/icde.2008.4497422.

[84] Yuan, G., Xia, S., Zhang, L., and Ji, C. Trajectory outlier detection algorithm based on structural features. *Journal of Computational Information Systems 7*, 11 (2011), 4137–4144.

[85] Zhang, D., Li, N., Zhou, Z.-H., Chen, C., Sun, L., and Li, S. iBAT: detecting anomalous taxi trajectories from GPS traces. In *UbiComp '11:*

[86] Gidófalvi, G., and Dong, F. When and where next: Individual mobility prediction. In *MobiGIS'12: Proc. First ACM SIGSPATIAL International Workshop on Mobile Geographic Information Systems* (2012), ACM Press, pp. 57–64. doi:10.1145/2442810.2442821.

[87] Jeung, H., Liu, Q., Shen, H. T., and Zhou, X. A hybrid prediction model for oving objects. In *Proc. 24th IEEE International Conference on Data Engineering* (2008), IEEE, pp. 70–79. doi:10.1109/icde.2008.4497415.

[88] Backstrom, L., Sun, E., and Marlow, C. Find me if you can: Improving geographical prediction with social and spatial proximity. In Proc. 19th International Conference on World Wide Web (2010), ACM Press, pp. 61–70. doi:10.1145/1772690.1772698.

[89] Froehlich, J., and Krumm, J. Route prediction from trip observations. In Society of Automotive Engineers (SAE) World Congress (2008), SAE International. doi:10.4271/2008-01-0201.

[90] Krumm, J. Where will they turn predicting turn proportions at intersections. Personal and Ubiquitous Computing 14, 7 (2009), 591–599. doi:10.1007/s00779-009-0248-

[91] Li, Z., Ding, B., Han, J., Kays, R., and Nye, P. Mining periodic behaviors for moving objects. In KDD '10: Proc. 16th ACM SIGKDD international conference on Knowledge discovery and data mining (2010), ACM Press, pp. 1099–1108. doi:10.1145/1835804.1835942.

[92] Jeung, H., Shen, H. T., and Zhou, X. Convoy queries in spatio-temporal databases. In ICDE '08: Proc. 2008 IEEE 24th International Conference on Data Engineering (2008), IEEE Computer Society, pp. 1457–1459. doi:10.1109/ICDE.2008.44

[93] Vieira, M. R., Bakalov, P., and Tsotras, V. J. On-line discovery of flock patterns in spatio-temporal data. In GIS '09: Proc. 17th ACM SIGSPATIAL International Conference on Advances in Geographic Information Systems (2009), ACM, pp. 286–295. doi:10.1145/1653771.1653812.

[94] Gao, P., Kupfer, J. A., Zhu, X., and Guo, D. Quantifying animal trajectories using spatial aggregation and sequence analysis – a case study of differentiating trajectories of multiple species. Geographical Analysis 48, 3 (2016), 275–291. doi:10.1111/gean.12098.

[95] Ying, J. J.-C., Lee, W.-C., Weng, T.-C., and Tseng, V. S. Semantic trajectory mining for location prediction. In GIS '11: Proc. 19th ACM SIGSPATIAL International Conference on Advances in Geographic Information Systems(2011), ACM, pp. 34–43. doi:10.1145/2093973.2093980.

[96] Anagnostopoulos, T., Anagnostopoulos, C., and Hadjiefthymiades, S. An online adaptive model for location prediction. Autonomic Computing and Communications Systems 23 (2010), 64–78. doi:10.1007/978-3-642-11482-3_5.

[97] Ying, J. J.-C., Lee, W.-C., and Tseng, V. S. Mining geographic-temporal-semantic patterns in trajectories for location prediction. ACM Transactions on Intelligent Systems and Technology 5, 1 (2013), 1–33. doi:10.1145/2542182.2542184.

[98] Monreale, A., Pinelli, F., Trasarti, R., and Giannotti, F. WhereNext: A location predictor on trajectory pattern mining. In KDD '09: Proc. 15th ACM SIGKDD international conference on Knowledge discovery and data mining (2009), ACM, pp. 637– 646. doi:10.1145/1557019.1557091.

[99] Li, X., Han, J., Kim, S., and Gonzalez, H. Roam: Rule- and motif-based anomaly detection in massive moving object data sets. In Proc. 2007 SIAM International Conference on Data Mining (2007), Society for Industrial and Applied Mathematics (SIAM), pp. 273–284. doi:10.1137/1.9781611972771.25.

[100] Renso, C., Baglioni, M., De Macedo, J. A. F., Trasarti, R., and Wachowicz, M. How you move reveals who you are: understanding human behavior by analyzing trajectory data. Knowledge and Information Systems 37, 2 (2012), 331–362. doi:10.1007/s10115-012-0511-z.

[101] Liu, Y., Wang, F., Xiao, Y., and Gao, S. Urban land uses and traffic 'sourcesink areas': Evidence from gps-enabled taxi data in shanghai. Landscape and Urban Planning 106, 1 (2012), 73–87. doi:10.1016/j.landurbplan.2012.02.012.

[102] Zheng, Y., Liu, Y., Yuan, J., and Xie, X. Urban computing with taxicabs. In UbiComp '11: Proc. 13th international conference on Ubiquitous computing (2011), ACM, pp. 89–98. doi:10.1145/2030112.2030126.

[103] Stevens, M., and D'Hondt, E. Crowdsourcing of pollution data using smartphones. In Proc. Workshop on Ubiquitous Crowdsourcing at UbiComp '10 (2010), ACM, pp. 1–3.

[104] Shang, J., Zheng, Y., Tong, W., Chang, E., and Yu, Y. Inferring gas consumption and pollution emission of vehicles throughout a city. In KDD '14: Proc. 20th ACM SIGKDD international conference on

Knowledge discovery and data mining (2014), ACM, pp. 1027–1036. doi:10.1145/2623330.2623653.

[105] Xiao, X., Zheng, Y., Luo, Q., and Xie, X. Inferring social ties between users with human location history. Journal of Ambient Intelligence and Humanized Computing 5, 1 (2012), 3–19. doi:10.1007/s12652-012-0117-z.

[106] Wu, F., Lei, T. K. H., Li, Z., and Han, J. Movemine 2.0: Mining object relationships from movement data. Proceedings of the VLDB Endowment 7, 13 (2014), 1613– 1616. doi:10.14778/2733004.2733043.

[107] Stein, M., Häussler, J., Jäckle, D., Janetzko, H., Schreck, T., and Keim, D. Visual soccer analytics: Understanding the characteristics of collective team movement based on feature-driven analysis and abstraction. ISPRS International Journal of Geo-Information 4, 4 (2015), 2159–2184. doi:10.3390/ijgi4042159.

[108] Laube, P., Dennis, T., Forer, P., and Walker, M. Movement beyond the snapshot: Dynamic analysis of geospatial lifelines. *Computers, Environment and Urban Systems 31*, 5 (2007), 481–501. doi:10.1016/j.compenvurbsys.2007.08.002.

[109] Gidofalvi, G., Huang, X., and Pedersen, T. B. Privacy-preserving data mining on moving object trajectories. In *2007 International Conference on Mobile Data Management* (2007), IEEE, pp. 60–68. doi:10.1109/mdm.2007.18.

4

An End–End Framework for Autonomous Driving Cars in a CARLA Simulator

Vedant Pandya, Shivanshu Shrivastava, Anuja Somthankar, and Megharani Patil

Department of Computer Engineering, Thakur College of Engineering and Technology, India
Email: pandyavedant18@gmail.com; shivanshu123shrivastav@gmail.com; anujasom@gmail.com; megharani.patil@thakureducation.org

Abstract

Autonomous vehicles are considered to be one of the most celebrated applications of artificial intelligence techniques. Self-driving cars are driven by complex deep learning neural networks. This book chapter proposes an end–end framework, which can make driving decisions, based on the surroundings, also accounting for the traffic light. Using images, which will be provided by a camera onboard, the framework will make predictions, which will allow it to efficiently and safely maneuver the vehicle. The framework consists of a traffic signal detection model, followed by a deep neural network which takes the input from traffic signal detection along with the images from the camera to generate the next navigation controls.

4.1 Introduction

As more research and development is done in the field of artificial intelligence and deep learning, the applications of the same are being used to make the lives of people around the world more comfortable. As time progresses, more and more people are becoming less prejudiced about having technology in their lives. In 1939, Norman Bel Geddes, at a general motor exhibit, showcased the world's first autonomous car, controlled by radio waves generated with metal spikes on the road, which was then upgraded to be controlled by currents flowing through the road. In the modern era, the availability of features like assisted braking and parking has helped turn cars into semi-autonomous. However, very few cars are completely autonomous. Rapid advancements in technology are required to facilitate the development of a fully autonomous vehicle.

The development of an autonomous vehicle consists of the following four components:

1. Sensors – To map the environment, sensors like radar and LiDAR are used.
2. Perception – Perception and localization read the information from sensors and estimate some relevant features representing the situation from it.
3. Planning – Motion planning selects the appropriate maneuver and trajectory of the vehicle.
4. Control – It is used for computing steering and acceleration commands to allow the vehicle to follow the trajectory.

4.2 Related Work

4.2.1 Autonomous driving simulators

In the coming future, self-driving cars will be ubiquitous. They can improve the safety of a passenger as well as reduce congestion on the roads. However, a huge obstacle to the progress of development of these vehicles is the immense cost to test them, to make sure that they are safe to be released. To train and test this software, millions of test cases must be performed, all of which may not be possible for even the most experienced developers. To accelerate this process, simulators play a huge role. They will help us design various test cases and scenarios and test our software for no cost and quickly. It also provides us with insights related to the motion of the vehicle. Some examples of such simulators are CARLA, SUMMIT, Udacity, etc. In this project, we shall be using the CARLA, or Car Learning to Act, to simulate a car and test our models. [Alexey Dosovitskiy *et al.*, 2017] introduced this simulator, which provided digital assets like vehicles, buildings, etc. This simulator provides flexible options ranging from sensors to changing weather conditions. It is an open-source simulator, built from scratch, which serves as a modular and flexible API to address a variety of obstacles that are found during testing. It was our preferred choice due to some of its features like scalability via a server multi-client architecture, autonomous driving sensor suite, and fast simulation for planning and control.

4.2.2 Object detection

Object detection refers to locating and identifying an entity in a given image. Object detection consists of feature extraction [Tiwari *et al.*, 2013], feature processing [Yan *et al.*, 2014], and object classification. Older object detection methods were based on simpler models and template matching processes [e.g., Fischler and Elschlager, 1973]. Techniques based on statistical classifiers (e.g., neural networks, Bayes, SVM, Adaboost, etc.) were presented [e.g., Fleuret and Geman, 2001; Osuna *et al.*, 1997]. These techniques laid the foundation for the research in the training and testing procedures of many machine learning techniques.

4.2.3 Literature review

Kichun *et al.* (2014) explain automation of cars in five basic steps which are perception, localization planning, control, and system management using technologies like LiDAR, radar, and vision. Sorin *et al.* (2020) grouped multiple deep learning autonomous driving systems and introduced deep learning methods for visual localization while comparing the two most popular sensors, LiDAR and vision. The authors also listed down various dataset sources for training self-driving systems. In [Danilo and Guilherme, 2012], a novel integration of planner, perception, and controller was provided to create a complete navigation system. Implementation and evaluation techniques were also provided in form of vector-fields.

Daniel (2017) highlights the importance of hardware and its crucial role in any autonomous car and summarizes a few areas where semiconductors can help in improving autonomous driving capabilities, by presenting key hardware technologies like comparators, compute, and accelerators. Chandravee *et al.* (2017) present a novel idea for semi-automatic cars which can be trained while driving, which can lead to better generalization of the model, and also introduce an overall driving score, which helps in ensuring security and safety.

In [Arnab Jana *et al.*, 2019], a survey was conducted which mapped the socio-economic standards of the respondents. The survey was made comparable with international studies. The model forecasts a person's mindset toward autonomous vehicles (AVs) as four degrees of interest, not at all interested, slightly interested, moderately interested, and very interested, and draws its relationship with the socio-economic characteristics of users such as age, gender, employment, and education levels on one hand and their desired activities, perceived concerns, and benefits of AV, etc., on the other hand.

MNL model offered a global statistic, i.e., one single value explaining the relationship between public interest levels and other independent variables considered in the model. It explains how the degrees of a person's interest in autonomous vehicles depend on other input variables.

Stéphanie Lefèvre *et al.*, (2016) combine two methods: learning-by-demonstration, which can generate commands that feel natural to the passengers, and predictive control, which relies on model-based predictions to make decisions and provides safety and stability. The driver model either learns continuously or is replaced, without having to readjust any module. The confidence value used gives a weight to each maneuver done in the learning by demonstration phase. By putting a set of limits on the vehicle state, the MPC ensures safety.

Kichun Jo, Yongwoo Jo, Jae *et al.* (2015) demonstrate a precise localization system based on the combination of data from onboard motion sensors in automobiles, a low-cost GPS, cameras, and a detailed digital map. The authors summarize research related to vehicle localization. It presented the probabilistic noise modeling of the RSM features using multiple cameras and described the fundamental flow of the particle-filter-based localization algorithm.

Two authors, John Leonard Jonathan How and Seth Teller (2008), thought of a perception-driven autonomous urban vehicle, which included many heterogeneous sensors with significant communications and computation bandwidth to capture and process high-resolution, high-rate sensor data. It included the following sub-systems: road paint detector, lane tracker, obstacle detector, hazard detector, fast vehicle detector, positioning module, navigator, drivability map, motion planner, and controller. A robust and flexible software architecture based on a novel lightweight UDP Ethernet protocol message-passing system supports these modules. DeepPicar, a low-cost deep neural network based autonomous car, was developed by Michael G. Bechtel *et al.* (2018). It uses a very computationally effective model, has an embedding computing platform for real-time inferencing, and has a very cost-effective model for implementation. The training resources for this were a lot and require a lot of investment, which brought the uniform training time down to 25 ms.

4.3 Proposed Work

4.3.1 Methodology

The proposed framework consists of a traffic signal detection model which detects and classifies traffic signals in images and a deep neural network which takes images and traffic signal detection results to predict the steering angle, brakes, and throttle for the next frame. Real-time images from the onboard camera are fed to a CNN, which detects the traffic light in the images and, consequently, labels them as red or green. This information along with the RGB image is then fed to a neural network, which then generates the next decision. The tools required for the project are as follows:

1. Python
2. Numpy and Pandas for data cleaning
3. Matplotlib for data visualization
4. Keras for model building
5. Google Colab, Visual Studio Code, and Spyder as IDE
6. Python flask for HTTP server
7. CARLA Simulator

4.3.2 Work flow

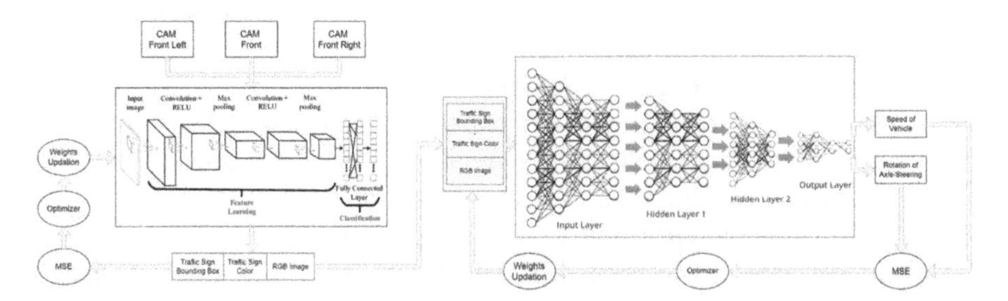

Figure 4.1 Flow of work.

4.3.3 Traffic signal detection

Traffic signal detection consists of predicting the category and the coordinates of the traffic signals that vehicles encounter, inside the CARLA simulator, using CARLA object detection dataset. The following steps are followed to implement the prediction of object detection:

1. Dataset selection
2. Dataset loading
3. Model selection, training, and inference

Table 4.1 Columns of CARLA object detection dataset.

Column name	Description
Name	Class of object
Pose	Whether the object is rotated
Difficult	Whether the object is difficult to detect
Truncated	Whether the object is truncated
Bounding box	Xmin, Ymin, Xmax, and Ymax of bounding box

4.3.3.1 Dataset selection

Traffic signal detection requires a dataset that contains images along with bounding boxes over the detected traffic signals and the traffic signal color, viz. red and green. For this purpose, the CARLA object detection dataset was used. The dataset contains 1028 images. As shown in Figure 4.1, the dataset has images with the data of bounding boxes which are stored in VOC format, i.e., XML files with the coordinates and the class for the object from "vehicle," "traffic-light," "traffic-sign," "bike," and "motorbike."

4.3.3.2 Dataset pre-processing and loading

The Udacity Sim dataset consists of 1028 images, each containing bounding boxes for objects like traffic light, vehicles, etc. Since our model only needs traffic signals, only the bounding boxes of traffic signals were retained, and all other objects were removed from the dataset.

As seen in Figure 4.2, out of the total traffic light instances in the dataset, 442 instances are of red light and the remaining 264 are of green light [Jean de Mello, Lucas Tabelini *et al.*, 2020]. The annotations were decoded from XML and mapped to each image. All of the images were resized to 448 × 448, and the dataset was split into training and test, with the test dataset accounting for 20% of the dataset, resulting in several training images being 820 and the number of test images as 105. To avoid a RAM crash, a generator object was created which loads images in RAM in batches and discards them once those images have been used for training. This process helps to train models on datasets whose size exceeds the RAM. Since only traffic lights are to be detected, there are two classes here, viz. "traffic-light" and "other."

4.3.3.3 Model selection

YOLO [Juan Du, 2018] is an algorithm that uses neural networks to deliver object detection in real time. This algorithm is prevalent because it gives good accuracy and is efficient and fast. After the definition of the model, the training of the model was done in batches of 4 for 20 epochs.

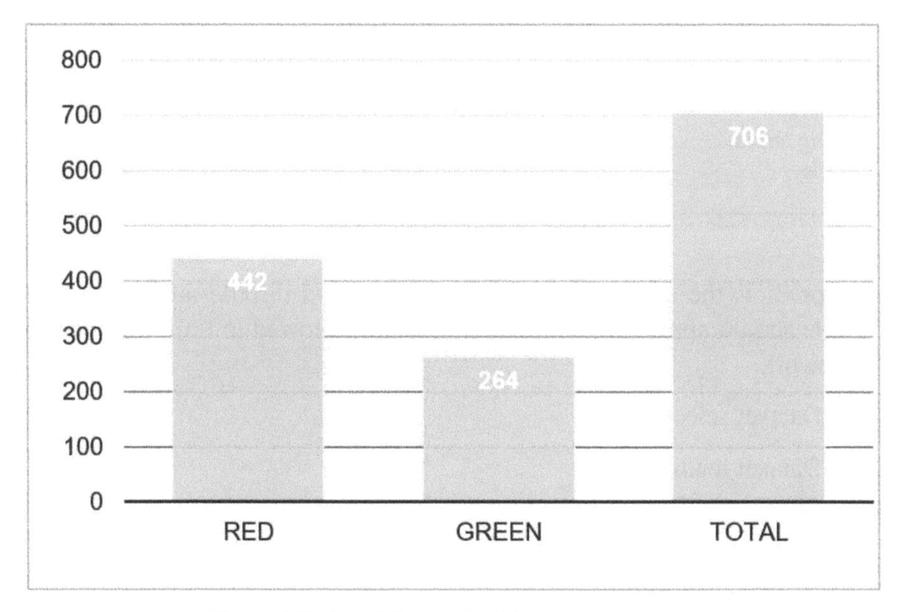

Figure 4.2 Breakdown of training and test images.

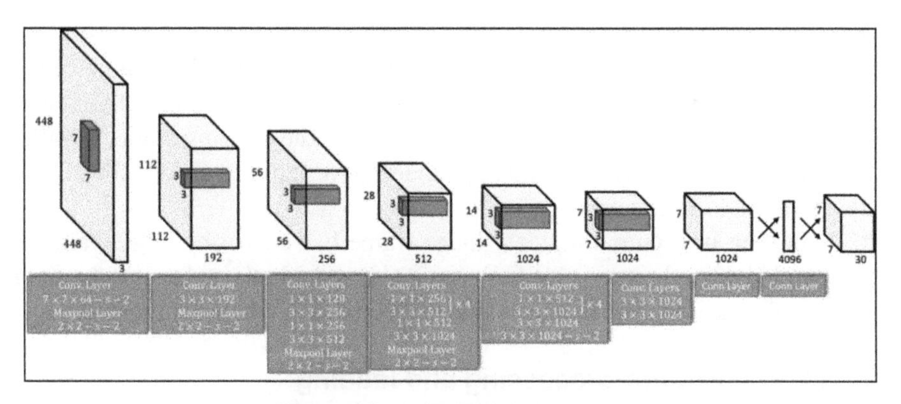

Figure 4.3 YOLO V3 architecture.

A custom loss function was devised to calculate loss, as a YOLO pipeline performs regression and classification. Adam optimizer has been used for changing weights after each step of gradient descent.

4.3.4 End–End framework

The proposed framework performs traffic signal detection to identify traffic light signals in a given image, identifies the color of the traffic light, and

Table 4.2 Columns of dataset.

Column name	Description
Image	Stores the path of the image
Steering angle	The steering angle of the vehicle at the moment
Throttle	Value of throttle at the moment
Brake	Value of brake at the moment

then predicts the steering angle, brake value, and throttle amount that the vehicle should apply. The following steps are followed to train the end–end framework:

1. Dataset selection
2. Dataset loading
3. Model

4.3.4.1 Dataset selection

For creating the end–end framework, a dataset was required which would give good results when it was simulated using the CARLA simulator [Alexey Dosovitskiy, German Rosc *et al.*, 2017]. As no functional dataset existed for the same, extraction of images was done through the auto-pilot feature of the simulator. To make the training more generalized, images from all five towns are captured. At each point an image is captured, the steering angle, throttle, and brake are also stored. This data is then converted into a CSV file. There are a total of 1725 images.

As seen in Figure 4.4, each image frame has a steering angle value in between [–1,1], a throttle value between [0,1], and a brake value is either 0 or 1. It can also be observed that the value of throttle is 0, whenever the brake is 1.

4.3.4.2 Dataset pre-processing and loading

In dataset pre-processing, the traffic light signal detection model trained before is applied on the images to get the area of the bounding box over the detected traffic signal. A mask is then applied to identify the color of the signal. After this, two columns containing this data are appended to the original dataset, as seen in Figure 4.5. The images were then converted from BGR to RGB format and resized to 150 × 150. The target variables are throttle, brake, and steering angle, which will be predicted using the area, signal color, and image. The data was split into train and test sets, with a ratio of 4:1 split.

As seen in Figure 4.5, the column of area denotes the area that the traffic light covers in the image frame. This metric was added so that the model can

image_name	steering_angle	throttle	brake
_out/003430.png	0.349534959	0.699999988	0
_out/003431.png	0.350039244	0.699999988	0
_out/003442.png	0.350504726	0.699999988	0
_out/003455.png	0.351008981	0.699999988	0
_out/003468.png	0.351474464	0.699999988	0
_out/003480.png	0.351978719	0.699999988	0
_out/003493.png	0.047408	0.699999988	0
_out/003505.png	0.005419074	0.502455294	0
_out/003517.png	-0.00473375	0.422162324	0
_out/003529.png	-0.025265656	0.395808518	0
_out/003541.png	-0.027244586	0.386252284	0
_out/003553.png	-0.019383475	0.382840812	0
_out/003565.png	-0.016234063	0.381289989	0
_out/003576.png	-0.00852174	0.380322933	0
_out/003589.png	-0.008803172	0	1
_out/003600.png	-0.010702723	0	1
_out/003613.png	-0.010173735	0	1
_out/003625.png	-0.010165093	0	1
_out/003636.png	-0.010167775	0	1
_out/003648.png	-0.010167775	0	1
_out/003660.png	-0.010167775	0	1
_out/003672.png	-0.010167775	0	1

Figure 4.4 Snippet of dataset.

learn to choose the closest traffic signal when more than one traffic light is present in the picture. The signal is a categorical variable whose value is 1 if the detected traffic light color is red and 0 otherwise.

4.3.4.3 Model

As stated earlier, the end–end framework consists of taking outputs from the traffic detection model along with the RGB image and predicting the vehicle controls. The end–end framework is a multi-input model, where the first input is the RGB image of size 150 × 150 and the second input is an array of [area, traffic_signal].

As seen in Figure 4.6, the RGB image of size 150 × 150 is fed to a VGG16 model. The VGG16 model is loaded with pre-trained weights and all its layers are frozen. The output from VGG16 is flattened and then fed to a dense layer of size 62. The output from traffic signal detection model is then fed to a different input layer of size 2. Both of the layers are concatenated and fed to a dense layer of size 64, followed by a dense layer of 32,

	image_nar	steering_a	throttle	brake	area	signal
0	_out/0034	0.349535	0.7	0	0	0
1	_out/0034	0.350039	0.7	0	0	0
2	_out/0034	0.350505	0.7	0	0	0
3	_out/0034	0.351009	0.7	0	0	0
4	_out/0034	0.351474	0.7	0	0	0
5	_out/0034	0.351979	0.7	0	0	0
6	_out/0034	0.047408	0.7	0	0	0
7	_out/0035	0.005419	0.502455	0	0	0
8	_out/0035	-0.00473	0.422162	0	0	0
9	_out/0035	-0.02527	0.395809	0	0	0
10	_out/0035	-0.02724	0.386252	0	0	0
11	_out/0035	-0.01938	0.382841	0	0	0
12	_out/0035	-0.01623	0.38129	0	0	0
13	_out/0035	-0.00852	0.380323	0	0	0
14	_out/0035	-0.0088	0	1	0	0
15	_out/0036	-0.0107	0	1	0	0
16	_out/0036	-0.01017	0	1	833	1
17	_out/0036	-0.01017	0	1	988	1
18	_out/0036	-0.01017	0	1	936	1
19	_out/0036	-0.01017	0	1	936	1
20	_out/0036	-0.01017	0	1	936	1
21	_out/0036	-0.01017	0	1	936	1
22	_out/0036	-0.01017	0	1	936	1
23	_out/0036	-0.01017	0	1	936	1

Figure 4.5 Dataset after pre-processing.

followed by two dense layers of size 16, and the final layer of size 3 acts as output layer. As the outputs are of mixed type, two outputs are of numerical type (steering angle and throttle), and one output is categorical (brake), a custom loss function was defined. After the definition of the model, the training of the model was done in batches of 64. The custom loss function returns the sum of MSE of throttle and steering angle and the categorical cross-entropy of the brake. The entire set was trained for 100 epochs with Adam optimizer.

4.4 Evaluation Metrics

Choosing an appropriate evaluation metric is very important for creating robust deep learning models. At times, it is not possible to use built-in loss functions, and it is necessary to create custom functions that can help to better train the models. Table 4.3 mentions the different types of losses used in the models trained above.

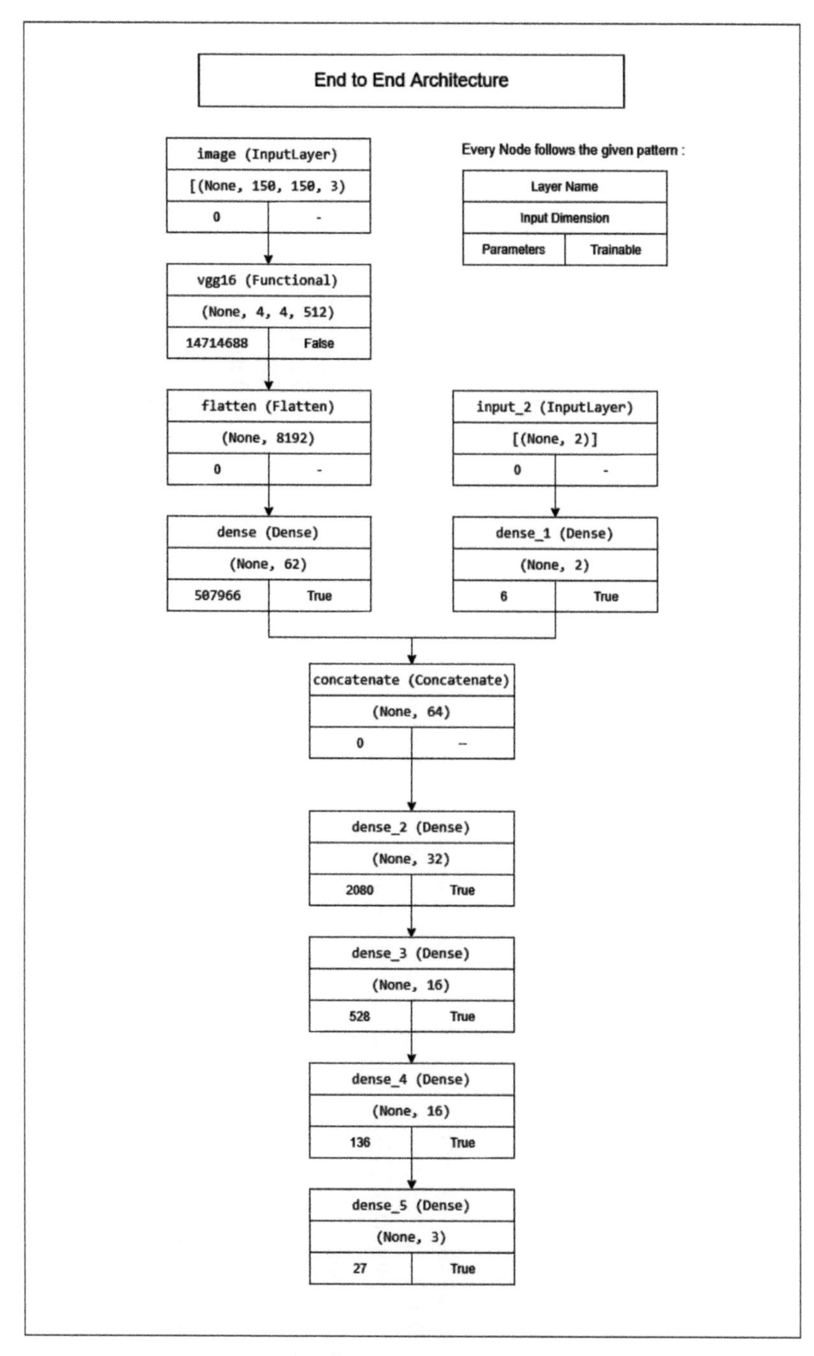

Figure 4.6 End–end model.

Table 4.3 Evaluation metrics.

Metric	Description	Formula
Mean squared error (MSE) [Cort Willmott *et al.*, 2005]	It is generally used in a regression function, to check how close the regression line to the dataset points is.	$MSE = \frac{1}{n}\sum_{1=1}^{n}\left(y_i - y_i^{\sim}\right)^2$
Root mean squared error (RMSE) [Cort Willmott *et al.*, 2005]	Often referred to as root mean squared deviation, its purpose is to find any errors in the numerical predictive models.	$RMSE = \sqrt{\sum \frac{(Y_{pred} - Y_{ref})^2}{N}}$
YOLO LOSS [J. Tao, H. Wang *et al.*, 2017]	The entire YOLO loss is the one that is followed in YOLO (You Only Look Once) algorithm; it is divided into four terms with each term having individual significance.	YOLO Loss Term 1 + YOLO Loss Term 2 + YOLO Loss Term 3 + YOLO Loss Term 4
YOLO Loss Term 1	It computes the loss that is related to the predicted bounding box position. It is one of the objects present in (*i,j*) the bounding box.	$\lambda_{coord}\sum_{i=0}^{s^2}\sum_{j=0}^{B} 1_{ij}^{obj}[(x_i - \hat{x}_i)^2 (y_i - \hat{y}_i)^2]$
YOLO Loss Term 2	It computes the loss that is related to the predicted bounding box width/height.	$\lambda_{coord}\sum_{i=0}^{s^2}\sum_{j=0}^{B} 1_{ij}^{obj}[(\sqrt{w_i} - \sqrt{\hat{w}_i})^2 (\sqrt{h_i} - \sqrt{\hat{h}_i})^2]$

YOLO Loss Term 3	It computes the loss related to the confidence score of all the bounding boxes calculated.	$\sum_{i=0}^{s^2}\sum_{j=0}^{B} 1_{ij}^{obj}(c_i - \hat{c}_i)^2 + \lambda_{noobj}\sum_{i=0}^{s^2}\sum_{j=0}^{B} 1_{ij}^{noobj}(c_i - \hat{c}_i)^2$
YOLO Loss Term 4	As we penalize the error when no object is present, it computes the sum of squared error of classification except for the 1^{obj} term.	$+\sum_{i=0}^{s^2} 1_i^{obj} \sum_{c\in classes} (p_i(c) - \hat{p}_i(c))^2$
Custom loss for end-to-end framework	The final prediction consists of a categorical term, i.e., brake, and two numerical values that are throttle and steering angle change; so we need to make a custom loss giving equal weight to all the losses.	MSE + CCE
Categorical cross-entropy (CCE) [C. H. Chen, P. H. Li *et al.*, 2020]	It is used in the classification task; it computes the loss concerning the class predicted and expected.	$CCE = (y \times \log(p) + (1 - y) \times \log(1 - p))$

4.5 Result of Prediction

4.5.1 Traffic signal detection

After training the model for 20 epochs, it can be seen that the validation loss starts to plateau after epoch 10. Hence, the predictions were made using the model trained till epoch 10, in the end–end framework. Figure 4.8 shows the result of the object detection method on an image retrieved from a front camera mounted on a car in CARLA.

The model can detect the bounding boxes of the traffic light and also provide a confidence score about each prediction. The area is calculated using bounding boxes detected and the traffic light is cropped from the image. A mask is then applied to the cropped traffic light image, to get its color.

4.5.2 End–End framework

As seen in Figure 4.9, the training loss and the validation loss plateaued after the 10th epoch. Hence, the model trained till epoch 10 was chosen for prediction. After training, the trained model is saved in .h5 format. The trained model is then loaded and a continuous image feed from the CARLA simulator

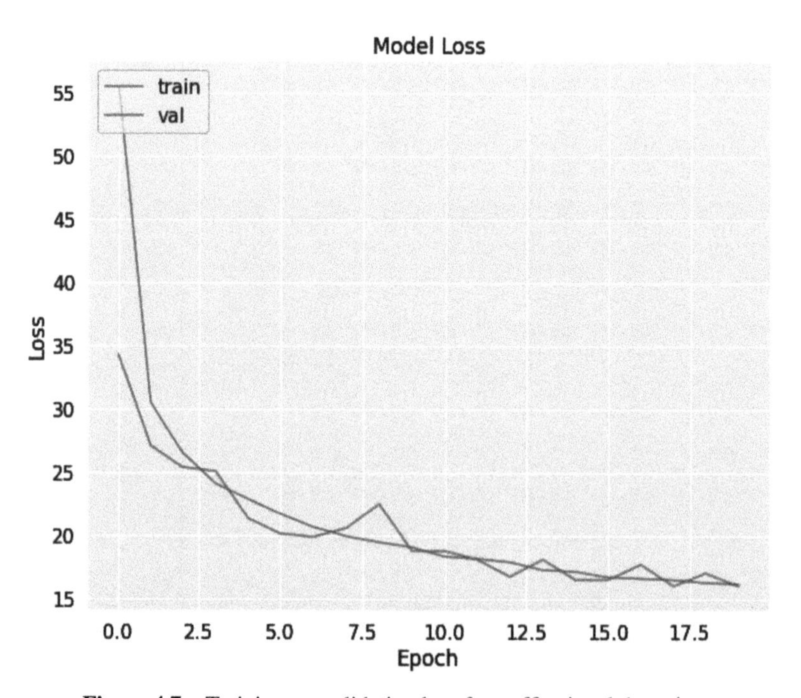

Figure 4.7 Training vs. validation loss for traffic signal detection.

Figure 4.8 Traffic signal detection.

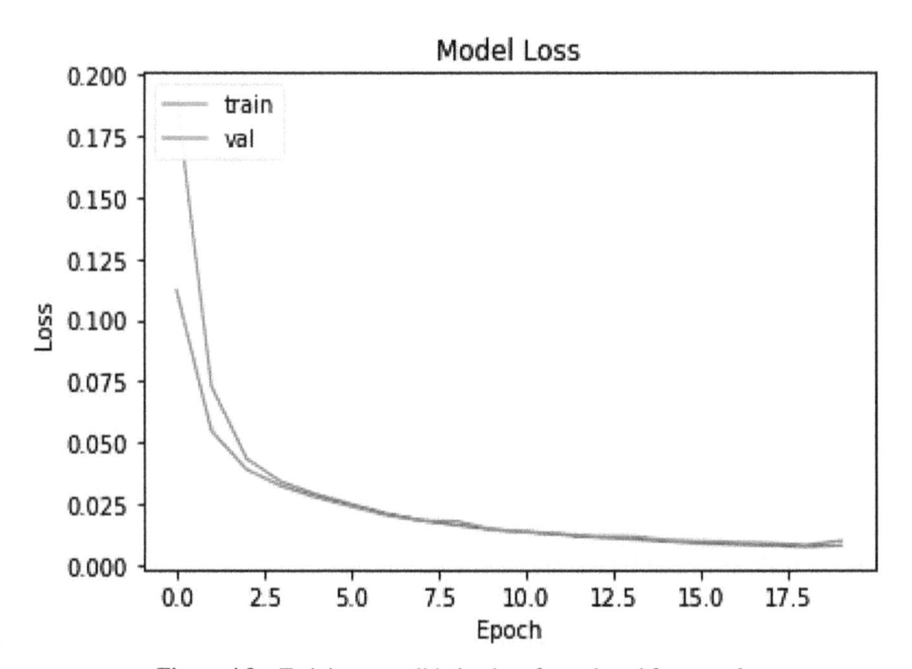

Figure 4.9 Training vs. validation loss for end–end framework.

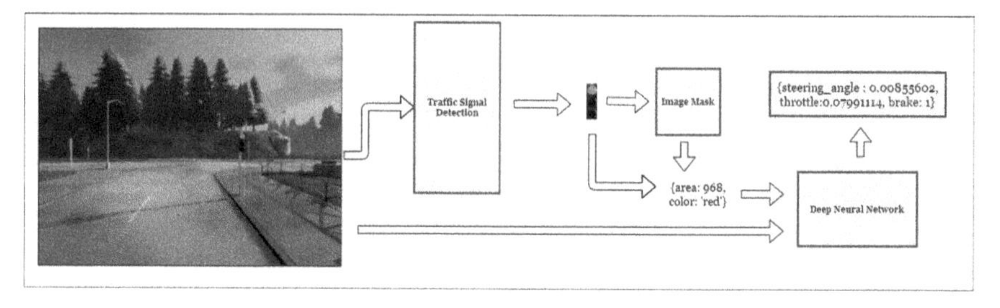

Figure 4.10 End–End framework result.

is fed to the model, which, in turn, controls the vehicle. The model sets the steering angle, throttle, and brakes, based on the inputs it received from the traffic signal detection model and the RGB image. To evaluate the model performance, the model is deployed on various town maps available in CARLA simulator, where it runs successfully.

Consider an example as shown in Figure 4.10; the model first detects the traffic signal and identifies it as red color. It then passes the area of a traffic signal, the traffic light color, and the original RGB image to the neural network which then predicts the steering angle, throttle, and brake. It can be observed that it applied brakes at a red signal and kept throttle to be zero, and steering angle toward the right, as the approaching turn is a right turn.

4.6 Conclusion

An end–end framework was proposed for autonomous driving, which processes a given image, detects traffic signals in the given image, identifies the color of the traffic lights, and then, using the area of a traffic signal, traffic light color, and the RGB image, predicts the steering angle, throttle and brake values. A traffic light detection model was trained and tested. A multi-input model that takes traffic light detection results and the image as input was trained on custom loss function on a dataset collected manually using CARLA simulator. Inferences made by observing the results show that the end–end framework is capable of driving an autonomous vehicle in the CARLA simulator while following the traffic signals. The model can be improved by training the detection model to detect other vehicles and pedestrians and collecting more data and training on larger datasets.

References

[1] Tiwari, Aastha, Anil Kumar Goswami, and Mansi Saraswat. "Feature extraction for object recognition and image classification." *International*

Journal of Engineering Research & Technology (IJERT) 2.10 (2013): 2278–0181.

[2] Yan, Junjie, et al. "The fastest deformable part model for object detection." *Proceedings of the IEEE Conference on Computer Vision and Pattern Recognition*. 2014.

[3] Fischler, M., and R. Elschlager. "The representation and matching of pictorial structures. 22 (1): 67–92." (1973).

[4] Fleuret, Francois, and Donald Geman. "Coarse-to-fine face detection." *International Journal of computer vision* 41.1 (2001): 85–107.

[5] Osuna, Edgar, Robert Freund, and Federico Girosit. "Training support vector machines: an application to face detection." *Proceedings of IEEE computer society conference on computer vision and pattern recognition*. IEEE, 1997.

[6] Jo, Kitchen, et al. "Development of autonomous car—Part I: Distributed system architecture and development process." *IEEE Transactions on Industrial Electronics* 61.12 (2014): 7131–7140.

[7] Grigorescu, Sorin, et al. "A survey of deep learning techniques for autonomous driving." *Journal of Field Robotics* 37.3 (2020): 362–386.

[8] De Lima, Danilo Alves, and Guilherme Augusto Silva Pereira. "Navigation of an autonomous car using vector fields and the dynamic window approach." *Journal of Control, Automation and Electrical Systems* 24.1 (2013): 106–116.

[9] Rosenband, Daniel L. "Inside Waymo's self-driving car: My favorite transistors." *2017 Symposium on VLSI Circuits*. IEEE, 2017.

[10] Basu, Chandrayee, et al. "Do you want your autonomous car to drive like you?." *2017 12th ACM/IEEE International Conference on Human-Robot Interaction (HRI*. IEEE, 2017.

[11] JANA, Arnab, et al. "Autonomous Vehicle as a Future Mode of Transport in India: Analyzing the Perception, Opportunities, and Hurdles." *Proceedings of the Eastern Asia Society for Transportation Studies*. Vol. 12. 2019.

[12] Lefevre, Stéphanie, Ashwin Carvalho, and Francesco Borrelli. "A learning-based framework for velocity control in autonomous driving." *IEEE Transactions on Automation Science and Engineering* 13.1 (2015): 32–42.

[13] Jo, Kichun, et al. "Precise localization of an autonomous car based on probabilistic noise models of road surface marker features using multiple cameras." *IEEE Transactions on Intelligent Transportation Systems* 16.6 (2015): 3377–3392.

[14] Leonard, John, et al. "A perception-driven autonomous urban vehicle." *Journal of Field Robotics* 25.10 (2008): 727-774.

[15] Bechtel, Michael G., et al. "Deeppicar: A low-cost deep neural network-based autonomous car." *2018 IEEE 24th international conference on embedded and real-time computing systems and applications (RTCSA)*. IEEE, 2018.

[16] de Mello, Jean Pablo Vieira, et al. "Deep traffic light detection by overlaying synthetic context on arbitrary natural images." *Computers & Graphics* 94 (2021): 76–86.

[17] Du, Juan. "Understanding of object detection based on CNN family and YOLO." *Journal of Physics: Conference Series*. Vol. 1004. No. 1. IOP Publishing, 2018.

[18] Dosovitskiy, Alexey, et al. "CARLA: An open urban driving simulator." *Conference on robot learning*. PMLR, 2017.

[19] Willmott, Cort J., and Kenji Matsuura. "Advantages of the mean absolute error (MAE) over the root mean square error (RMSE) in assessing average model performance." *Climate Research* 30.1 (2005): 79–82.

[20] Tao, Jing, et al. "An object detection system based on YOLO in traffic scene." *2017 6th International Conference on Computer Science and Network Technology (ICCSNT)*. IEEE, 2017.

[21] Chen, Chien-Hua, et al. "Robust Multi-Class Classification Using Linearly Scored Categorical Cross-Entropy." *202020 3rd IEEE International Conference on Knowledge Innovation and Invention (ICKII)*. IEEE, 2020.

5

IoT and Artificial Intelligence Techniques for Public Safety and Security

Vinod Mahor[1*], Sadhna Bijrothiya[2], Romil Rawat[3], Anil Kumar[4], Bhagwati Garg[5], and Kiran Pachlasiya[6]

[1]IES College of Technology, India
[2]PhD Maulana Azad National Institute of Technology, India
[3]Shri Vaishnav Vidyapeeth Vishwavidyalaya, Indore, India
[4]Government Engineering College, India
[5]Union Bank of India, India
[6]NRI Institute of Science and Technology, India
Email: [1*]vinodengg.mt@gmail.com; [2]sadhanaengg@gmail.com; [3]rawat.romil@gmail.com; [4]anil_kpawar@yahoo.com; [5]gargpratap@gmail.com; [6]pachlasiakiran@gmail.com

Abstract

The Internet of Things (IoT) is transforming how organizations and industries communicate and conduct their everyday operations. Its application has proven to be ideal for industries that manage large amounts of assets and coordinate complex and dispersed operations. An expert system must be able to defend and explain a judgment to a user in AI. Experts in information security must be able to explain why a system is secure to the general public. Artificial intelligence is also expected to play an important role in ensuring long-term public safety and security for growth. The different emerging sectors of IoT and traditional towns have been transformed into high-tech smart cities due to AI. The emergence of smart cities has just one goal: to raise people's living standards by incorporating technology into their daily lives. This chapter discusses a number of significant technologies as well as solutions to problems that citizens face as a result of a lack of digitalization. It addresses concerns such as public infrastructure, public safety, and security and provides ideal solutions. It focuses not only on AI but also on IoT, machine learning, deep learning, pattern reorganization, and big data analytics for the development of a smart city that is completely functional. The analysis identifies the major barriers to widespread adoption and proposes a research path to address each of them in a cost-effective manner. IoT for defense and public safety.

5.1 Introduction

The notion of a smart city is based on saving money, improved living standards, resource conservation, technological integration, and quicker transactions in every sector. It uses every part of technology to turn a complex infrastructure into a digital, sophisticated, and easier way of living. The word "technology" refers to the most prominent and growing disciplines, such as AI and the Internet of Things. Smart city is one that data is collected using a variety of technological sensors and devices. Its data is then utilized to improve operational processes by gaining insights on how to manage assets, resources, and services more efficiently. Data from people, devices, buildings, and assets is recorded and analyzed to measure and control urban transportation systems, power plants, utility companies, ground water connectivity, rubbish, crime recognition, information systems, schools, colleges, library services, medical facilities, and other community services [1]. Smart cities combine modern information and communication technologies (ICT) with a number of physical devices to the Internet of Things (IoT) network to improve local operations and services while also connecting residents. A smart city's major objective is to enhance policy efficiency, eliminate waste, alleviate unpleasant situations, increase economic and social quality, and raise the standard of living [2].

They, like smart cities, are the future. Both of these fields have a large radius, making it very hard to reach their ends. IoT provides the basis for linking every one of these connected technologies to build a grid, whereas AI intends to focus on combining technologies with even the most basic of objects. Many individuals have various ideas about what it means to live in a "smart city" [3]. This is explained in detail. It includes smart economic perspectives, smart government, new organizational methods, smart transportation, intelligent transportation, information sharing, strategy, achievement, and more efficient, easier, and qualified occupations [4].

The government has instituted numerous procedures to emphasize the necessity of a technology infrastructure in a city, and smart cities are the next big thing. The Ministry of Urban and Housing Affairs' "Smart Cities Mission" is one of the most visible tasks. It is indeed a platform where states

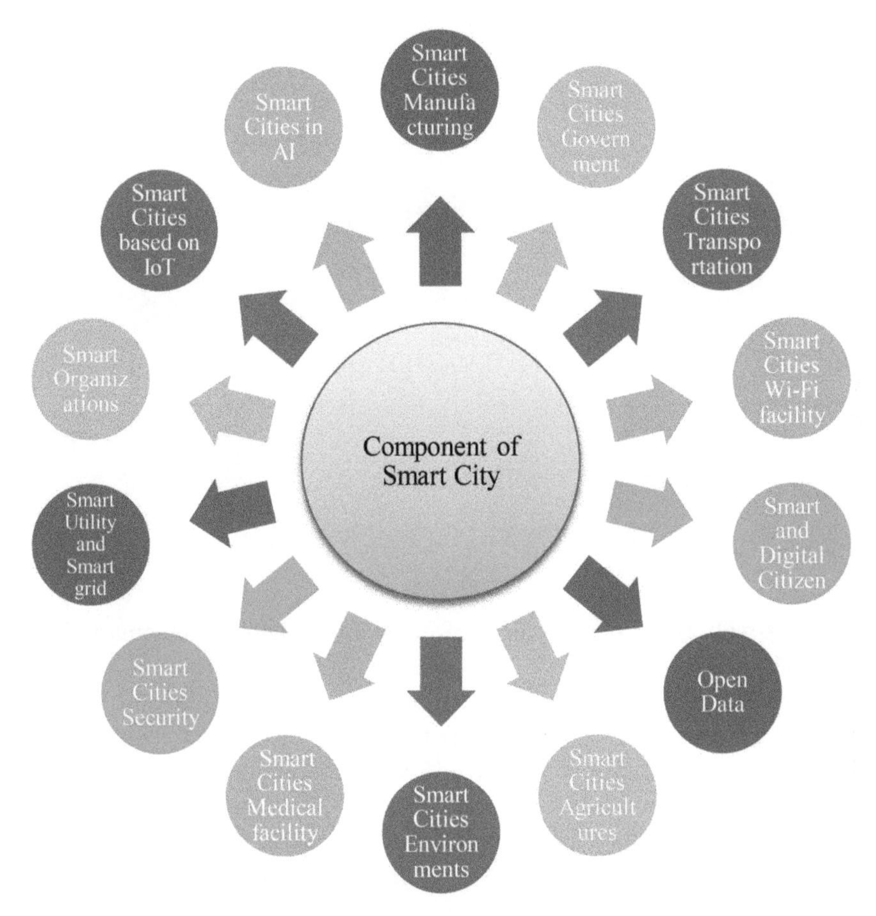

Figure 5.1 Components of smart cities.

may suggest cities to be transformed into smart cities. It is done like an antagonism, with cities being selected and rated according to their worth. Figure 5.2 depicts the many steps of the smart city selection process [5]. Every aspect to consider: the number of tasks completed in that city, the overall cost of those tasks, the overall cost of development in a certain region, every aspect to consider: chores are the entire pan-city solution, as well as the most harmed city population. Certain cities have provided certain essential infrastructures: a reasonable standard of living for their inhabitants, a healthy and long-term environment, and development that is open to all, and as a result of such an endearing notion, the notion of looking at compact regions and building a repeatable model that would function as a beacon for other ambitious cities [6]. It promotes the adoption of "smart" solutions to complex

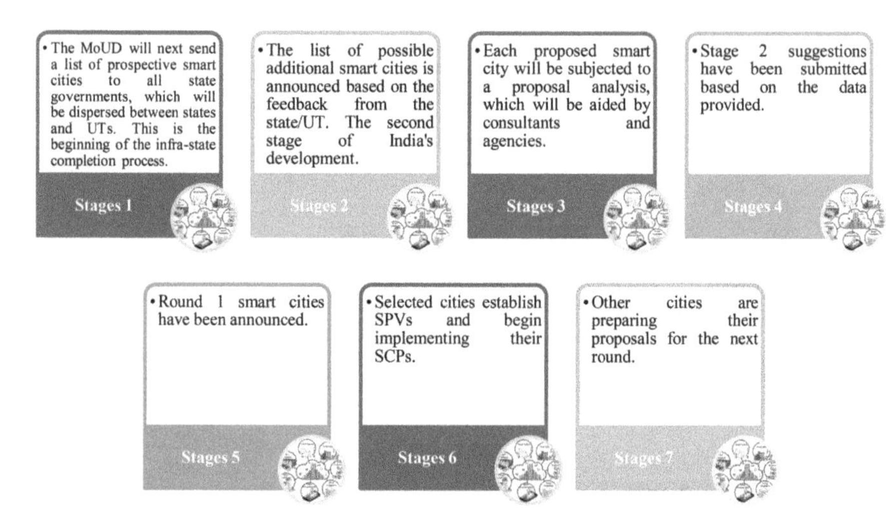

Figure 5.2 Stages of a smart city project.

issues, as well as the development of every component of the smart city, as illustrated in Figure 5.1.

The smart cities missions' goal intends use technology to generate financial growing and keep improving the quality of people's lives. The goal of area-based development intend refit and rehabilitate regions, especially slums, into better-planned districts, therefore enhancing the city's overall quality of life. This holistic development will contribute to improved livability and incomes for everybody, as well as job possibilities for the poor and disadvantaged, resulting in inclusive cities [5–7].

5.2 Proposed Method

The study of AI-based programming allows computers to replicate human activities by mimicking intelligence in them. AI's main aims are learning, reasoning, and perception. Medical diagnosis, robot centralism, electric training, finance, remote detecting, ophthalmic charm acknowledgment, computer visions, semantics web, practical actuality, image preprocessing, game theory, and other applications already use this idea. The development of smart cities may be made comprehensive [7]. The city's AI learns in what way people utilize it: AI pattern appreciation technology is often used to manage massive amounts of raw data, such as mass transit ticket sales, police reports, and traffic sensors as well as weather stations. To expand the supply of this raw data, around 1 billion cameras are anticipated to be placed throughout infrastructure and government properties by 2020. Humans can only actively watch a small percentage of cameras. This is where deep learning enters the picture. It can count cars and pedestrians, scan license plates and recognize aspects, and monitor vehicle speeds to create patterns. It can also use satellite data to tally up the number of vehicles in a parking garage or measure the quantity of traffic in any other particular location at any given moment [8]. The LPR system from VIMOC Technologies can verify if cars parked in a parking lot have the proper permits. In the case of power supply and management, AI may be utilized to create a system that can effectively interact and intelligence with social controllers in the event of an emergency [9].

There is a great deal of inconvenient traffic and locating a parking spot. Parking spaces are scarce in areas where artificial intelligence is improving infrastructure. According to research, drivers spend around 107 hours each year looking for a parking spot. Individuals, businesses, and government organizations may work together to solve this problem by sharing and collecting real-time data. Drivers can offer real-time traffic and accident information to assist people in saving time and improving their routes. Passengers find it much more convenient to travel by tractable buses rather than those for which they must wait a lengthy period. They may plan their activities properly and save time as a result. It is known as linked public transportation technology [10].

There is a great deal of inconvenient traffic and locating a parking spot. This technology allows the general population to communicate with buses and trains, letting them know when they will arrive and informing them in advance if there will be a delay. Waste management is an essential aspect of smart city development. SITA Finland is a self-thinking robot developed by ZEN Robotics. It is an intelligent device that employs artificial intelligence to make judgments on its own and handle challenges linked to construction and demolition trash [11]. Based on adaptive signal management technology, it enables traffic lights to change in real time based on data collected from various cameras and other companies that update their apps with the most current information about traffic conditions at various places all over the city.

In large cities, this can save more than 10% of the overall journey time, and in areas with out-of-date signal timings, it can save even more. It can reduce travel time by 50%. Many countries are implementing this technology because it can minimize traffic congestion expenses owing to productivity and wasted fuel [12]. In places such as San Antonio, San Diego, Los Angeles, and Bellevue, the advantages of this technology have been applied, measured, and proved successful. A smart city application designed to manage emergency situations and urban traffic in real time. As a result, a smart city may

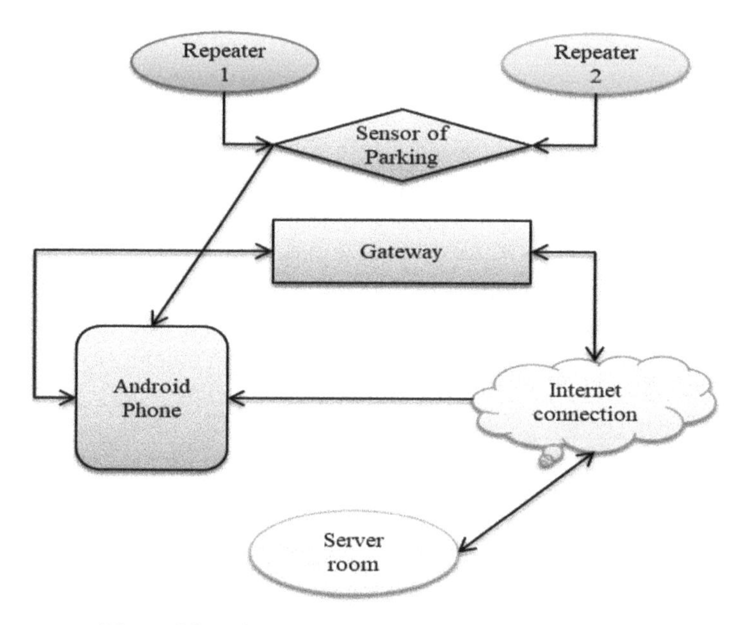

Figure 5.3 Diagram of the smart parking module's flow.

be better able to react to challenges than a city with a solely "transaction-oriented" relationship with its residents. The statement, however, is vague in its specifics, leaving it open to a number of perspectives. Sustainability is another essential component of smart cities [13]. Urbanization is expected to accelerate in the next few years. As per the United Nations, around 55% of the world's population now lives in a metropolitan or urban environment, with this percentage predicted to rise to 68% in the next decade.

According to some, smart cities are responsible for decreasing traffic congestion and not just saving fuel, but also saving lives and combating crime. Hunting for stolen cars and criminals is one means of protecting municipal safety. That time, the same LPR technology is employed [15]. Ambulance services and fire engines use sophisticated traffic signals to get to an urgent situation efficiently and smoothly. This city's vast data gathering helps to identify regions prone to frequent tragedies, determine their causes, and prevent them in the future. In the event of an accident, contact with the appropriate authorities and an automatic response management system can take control. In such circumstances, big data plays a crucial role in minimizing road deaths and prioritizing infrastructure upgrades.

Video surveillance, analytics, IoT-based drones, gunshot detection sensors, video surveillance, analytics, drones, and cyber security are several of the instruments that might be utilized to keep people safe in such a smart city [16]. Spark Cognition, which unveiled Deep Arm, an AI-powered antivirus, and Dark Trace, which aims to discover new cyber threats and anticipate better-quality information for guaranteeing security and safety, are two ground-breaking cyber security breakthroughs. A surveillance system based on artificial intelligence can track the transmission of microorganisms that cause various diseases. Smart health information systems that are stored in the cloud can be utilized to create more efficient treatment and healthcare suggestions and decisions [17, 18]. Citizens' input and involvement in the construction of a coherent environment might help smart cities become smart technologically.

5.3 Smart City Technology Framework

To offer public access to connected solutions, a variety of software is used in smart cities, user interfaces, network communication, as well as IoT and AI. IoT is the most important of these factors. IoT technology is an interconnected network of devices that communicate and share data. Automobiles, home appliances, as well as sensors, to mention a few, are examples of this. These gadgets capture data, which is then kept on the internet and networks, enabling both industry and government to be more efficient, both in terms of financial gain and improvement of people's lives on a daily basis [19].

To ensure security, other IoT devices use edge computing. Through the communication network, only the most critical and relevant information is provided. There is also a security and safety system in place to safeguard, monitor, and control data transmission from the smart city network. A quality and security system has also been installed to protect, monitor, and manage data transfer from the smart city network. In a smart city, technology implementation is important [20]. Figure 5.1 illustrates how smart city innovations are made up of multiple combinations of technological resources that combine to provide a wide range of smart city technologies with varied degrees of personal communication.

However, many of these issues may be addressed through the use of AI-enabled IoT. Residents' daily lives may be made more pleasant and safer by leveraging technological progress to support the new experience. The term "smart city" was coined as a result of this.

A smart city is one that uses information and technology to improve the quality and performance of urban services (such as energy and transportation), reducing resource consumption, prevent waste, and lower total costs. Smart cities not only have ICT but also use it in a way that benefits the people who live there.

- In a smart city, people and technology must be connected throughout the system architecture. Innovative services and communications infrastructure are two examples. A smart city is defined as "a network community that integrates broadband network connectivity"; "a flexible network community that integrates broadband network communication"; "service-oriented computer infrastructure built on open industry best practices"; and "innovative services to meet the expectations of governments and their employees, people, and business owners," by Hazapis and Yovanof.
- For instance, machine learning and artificial intelligence (AI) may be trained on data generated by smart city sensors to uncover trends. Computerized systems that analyze people's ongoing interactions with their urban environments can be used to assess the efficacy and effect of various policy initiatives.
- Access to public services is possible through every device connected in such an accessible city. Urban city is a logical extension of the digital city concept because it just likes the easy accessibility to all infrastructures.
- The physical component of IT systems is widely used in early-stage smart cities. A wired network is necessary to enable the IoT and wireless technologies that are at the heart of our increasingly connected lives. In a linked city, the public in general has access to continually updated physical and digital infrastructures. Telecommunication, cloud computing, IoT, and other linked technologies might then be used to boost human resource management and productivity.
- A hybrid city is one that combines a physical conurbation with a virtual city that is connected to it. It can be established by virtual design or by the presence of a critical mass of virtual members of the public in

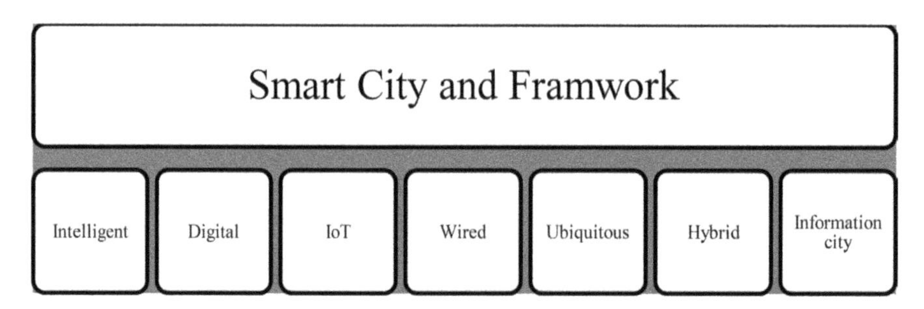

Figure 5.4 Smart cities framework with technologies.

a real-world metropolis. Heterogeneous space can assist in the implementation of future-state smart city services and integration efforts.

- The vast number of interactive devices in a smart city generates a massive amount of data. The way data is handled and stored is critical to smart city growth and security.

5.4 Other Technologies that are Connected

The growth of smart cities can be aided by many areas of ML and AI. Patterns recognizing: citizen-city synergy is a critical component of AI in smart cities. It assists in the study and analysis of current data in order to make better future decisions while dealing with significant changes in operations. Due to the complexity of cities, several problems may occur with the adoption of a clear model. The appropriate approach for each city must be determined in order to reduce the financial risk of significant upfront construction expenses. ML classification algorithms such as Bayes network (BN), naive Bayesian (NB), J48, and nearest neighbor may be used to forecast weather data, notably temperature and rain (NN) [21, 22].

Appreciation of images: This may be a fun way to beat the traffic, and it is already happening in New York City's congested streets. The NYC Department of Transportation has collaborated with IntelliScape.io to better analyze and mitigate significant traffic incidents in the city. This, if implemented in every congested city in the world, will have a positive impact on traffic reduction and, thus, save a significant amount of time [23].

The system can identify traffic bottlenecks, weather report, parking infractions, and deliver real-time notifications to local officials, thanks to a wonderful mix of ML and picture increase. The junction cameras gather and analyze activity before broadcasting real-time findings and actionable data, such as issuing traffic warnings to violating cars. For real-time analytics, the technique can be used in conjunction with weather, demographic, and location-specific information.

The MF Art of Mapping classifier can be used for object recognition in traffic surveillance systems. We produce a photograph of an unknown car, extract local features, submit the local feature descriptors to the MF Art Map classification, obtain a list of classes and the sample's membership values in these classes, and then use heuristics to produce results [24].

The Internet of Things (IoT): It is a system that integrates data from a large number of sensors. This information should be compressed to fit onto existing systems, verified for integrity and compliance with security requirements, and then processed and stored so that end users can access it [25].

Instrumented, networked, and intelligent are the three key characteristics of a smart city. This is accomplished using the Internet of Things, by setting up sensors (GPS, RFID, IR scanners using lasers, and so on), obtaining an in-depth understanding, connecting internet, location, tracking, monitoring, and management. It is capable of monitoring, operating, and administering devices remotely, as well as accessing and analyzing real-time data [26].

Figure 5.4 depicts some of the most important IoT applications. IoT may be used for effective water supply and usage, creative traffic congestion solutions, more dependable public transportation, energy-efficient building transformations, and public safety.

ICT is used in a variety of fields, including medical, transportation, tourism, government, crime prevention, disaster management, and so on. Ambient assisted living for older people is one of the uses. This technology allows patients to stay at home while having their bodily parameters monitored on a regular basis through the use of body sensors [27].

When used separately, artificial intelligence and IoT technologies are incompatible with one another; functioning in entirely different paradigms, yet merging the two might help a wide range of businesses. IoT and AI amalgamation are the coming together of two cutting-edge technologies that have the potential to open up a world of possibilities. Intelligent vehicles for self-driving cars may be the most promising option.

The concept is for a driverless automobile system, self-learning cars, and a smart city transportation system to use an artificially intelligent IoT. The idea is to employ an artificially intelligent IoT to create self-driving car systems, self-learning vehicles, and a smart city transport system. By reducing the number of people engaged, synchronizing data, and creating audit logs, IoT technology can help to streamline the renting process. Any intelligent car system will gain greatly from this. Following are a few examples [28].

- Using IoT-AI, it allows a diverse range of vehicles to communicate and share information.
- Cost savings, time saving, and human effort are decreased. To ensure that the learning process is modular, eliminate individual vehicle training to increase efficiency.

The major drivers of today's innovation are artificial intelligence (AI) and IoT. Both are expected to have a massive impact on our lives and generate trillions

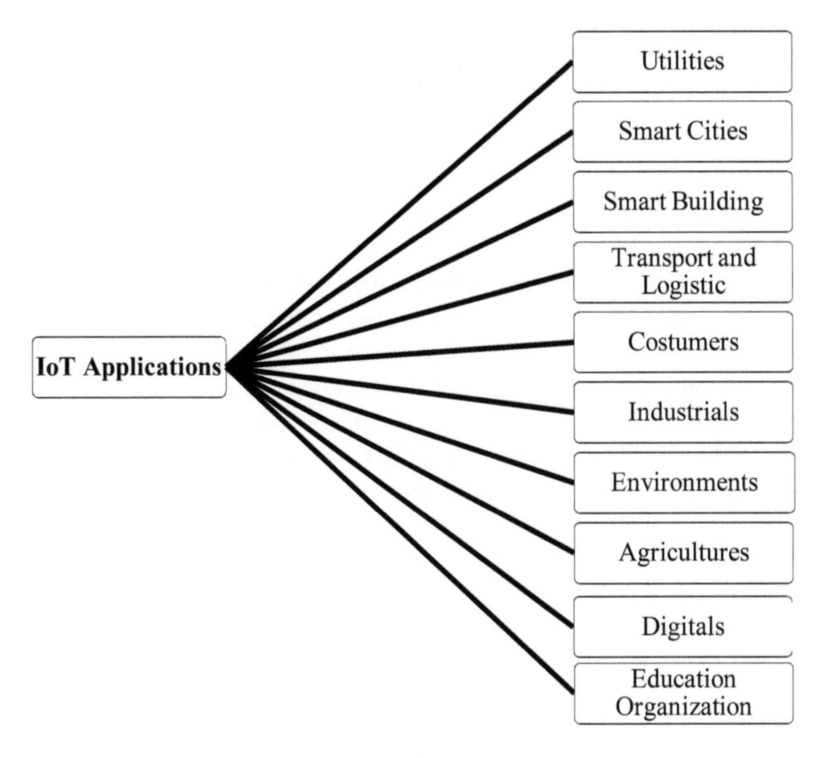

Figure 5.5 Internet of Things (IoT) applications.

of dollars for the international economy. On the one hand, the Internet of Things has issues with scalability, security, and efficiency. On the other hand, artificial intelligence has its own set of privacy and trustworthiness concerns. By complementing one another, the merging of these two technologies will revolutionize each other. The IoT can provide anonymity and trustworthiness, and AI can use the IoT to develop machine learning algorithms for security, scalability, and efficacy.

Big data: RFID chips on automobiles will aid in traffic regulation and measurement. Cisco is working on a novel method of producing electricity from waste collected in a city. This will help reduce the number of garbage trucks in the city. On sensor-based systems, wristbands may be worn by children playing in parks to track them if they become separated, making life safer. Advancement in this sector is through smart energy grids [28]. It could be used to detect the presence of people in a certain area and adjust the lighting system accordingly, allowing lightly populated areas to save electricity by dimming their lights. The resilience of a smart city to natural disasters may also be improved by using large amounts of data. Disaster managers can see potential threats, take actions to minimize them, and ultimately assist cities in being more resistant [29]. Big data collected from hospitals throughout the country may be used to identify common signs of a new disease and collaborate to raise awareness and discover a treatment; Thousands of people's lives might be saved. This method might be used on a global basis; doctors from many hospitals throughout the world are collaborating to eradicate such dreadful illnesses [30].

Deep learning: The approach was successfully used for the analysis of several data types, including video, pictures, audio, and textual data. In smart cities, a large amount of data in a time series is supplied by sensor devices. In recent years, deep learning has made major advances in AI, and it lends itself well to serial data acquisition. In smart cities, it will be used to estimate the quality of the air. In this model, the LSTM artificial neural network and SVR are employed [31, 32]. In recent years, deep learning has made tremendous advances in AI. It adapts nicely to the collection of genetic data. Along with water management and leak detection, energy savings, waste removal, and a variety of other tasks. Because once engines have been taught using deep learning approaches, the actuators that perform tasks automatically can be controlled [33].

5.5 Conclusion and Future Work

In this chapter, we discuss every term related to artificial intelligence: IoT-based smart cities, safety, and security. It delves deeply into each of the many approaches to smart city issues, as well as each of the potential answers. Deep learning, analytical thinking, picture recognition, big data analytics, and neural networks could be used to solve a variety of problems in traffic control, parking, hydropower, education, and the manufacturing sector. This chapter explains how the smart city sector is transforming as a result of the confluence of IoT and AI. It brings businesses, governments, and even countries together. Because of its decentralized nature and peer-to-peer qualities, IoT and AI technology are well-known and highly regarded. This is particularly important in light of predicted urban population growth, which would demand more efficient utilization of infrastructure and assets. The improvements will be made feasible through smart city services and applications, and this will improve people's quality of life.

In the future, the quality of our world's future cities will decide its fate; many are centuries old, while others are still under construction. There are roughly 650 cities on the globe, each having a population of over 550,000 people. Nevertheless, the top 25–30 cities currently account for half of the world's wealth. Over the next 20–25 years, cities' infrastructure spending is expected to be in the range of $35–$45 trillion. By 2020–2025, it is expected that over 45 cities in the world will be designated as smart cities. On top of that, the United Nations predicts that urban populations will nearly double by 2055, whereas the global population is rising from 7.2 billion to over 9.2 billion. There will be some stability and dispute as the standard of living, quality of life, and life expectancy improve.

References

[1] Ilhan Aydin, Mehmet Karakose, Ebru Karakose, "A Navigation and Reservation Based Smart Parking Platform Using Genetic Optimization for Smart Cities", 5th International Istanbul Smart Grid and Cities Congress and Fair (ICSG), Istanbul, Turkey, 2017

[2] Nomusa Dlodlo, Oscar Gcaba, Andrew Smith, "Internet of Things Technologies in Smart Cities", IST-Africa Week Conference, Durban, South Africa, 2016

[3] Thomas Bjørner, "How are smart cities perceived by project leaders and participants in an ongoing project: the challenge of evaluating smart cities", Smart City Symposium Prague (SCSP), Prague, Czech Republic, 2018

[4] Robert R Harmon, Enrique G Castro-Leon, Sandhiprakash Bhide, "Smart cities and the internet of things", Portland International Conference on Management og Engineering and Technology (PICMET), Portland, OR, USA. 2015

[5] Daniel Lorencik, Iveta Zolotova, "Object recognition in Traffic Monitoring Systems", World Syposium on Digital Intelligence for Systems and Machines", Kosice, Slovakia, 2018

[6] Ibrahim Kok, Mehmet Ulvi Simsek, Suat Ozdemir, "A deep learning model for air quality prediction in smart cities", IEEE International Conference on Big Data, Boston, MA, USA, 2017

[7] Chuanjie Yang, Guofeng Su, Jianguo Chen, "Using big data to enhance crisis response and disaster resilience for a smart city", IEEE 2nd International Conference on Big Data Analysis (ICBDA), Beijing, China, 2017

[8] B-IoT: Blockchain Technology for IoT in Intelligent Transportation Systems, http://iot.ed.ac.uk/projects/b-iot/, accessed on 2019.

[9] Tu, Y., Lin, Y., Wang, J., & Kim, J. U. (2018). Semi-supervised learning with generative adversarial networks on digital signal modulation classification. Comput. Mater. Continua, 55(2), 243–254.

[10] Zhou, S., Liang, W., Li, J., & Kim, J. U. Improved VGG model for road traffic sign recognition. CMC Comput., Mater. Continua, 57(1), 11–24, (2018).

[11] Z. Tang, X. Ding, Y. Zhong, L. Yang, and Li, K. A Self-Adaptive Bell–LaPadula Model Based on Model Training With Historical Access Logs. IEEE Transactions on Information Forensics and Security, Vol.13(8), pp. 2047–2061, 2018.

[12] D. Wang, L. Huang and L. Tang. "Synchronization criteria for discontinuous neural networks with mixed delays via functional differential inclusions", IEEE transactions on neural networks and learning systems, Vol. 29(5), pp. 1809–1821, 2017.

[13] Z. Cai, and L. Huang, Finite-time stabilization of delayed memristive neural networks: Discontinuous state-feedback and adaptive control approach. IEEE transactions on neural networks and learning systems, Vol. 29(4), pp. 856–868, 2017.

[14] D. Zeng, Y. Dai, F. Li, R. S. Sherratt, and J. Wang, Adversarial learning for distant supervised relation extraction. Computers, Materials & Continua, Vol. 55(1), pp. 121–136, 2018.

[15] K. Sgantzos, and I. Grigg, "Artificial Intelligence Implementations on the Blockchain: Use Cases and Future Applications," Future Internet, Vol. 11, no. 8, pp.170–182, 2019

[16] J. Sun, J. Yan, and K. ZK. Zhang, "Blockchain-based sharing services: What blockchain technology can contribute to smart cities," Financial Innovation, Vol. 2, no. 1, 26–35, 2016

[17] R. Shrestha, and S. Y. Nam, "Regional blockchain for vehicular networks to prevent 51% attacks," IEEE Access, Vol. 7, pp. 95021–95033, 2019

[18] Swetha Srivastava, Aditya Bisht, Neetu Narayan, "Safety and security in smart cities using artificial intelligence", 7th International Conference on Cloud Computing, Data Science & Engineering – Confluence, Noida, India, 2017

[19] Shubham Mathur, Uma Shankar Modani, "Smart City- a gateway for artificial intelligence in India", IEEE Students' Conference on Electrical, Electronics and Computer Science (SCEECS), Bhopal, India, 2016

[20] Jeannette Chin, Vic Callaghan, Ivan Lam, "Understanding and personalizing smart city services using machine learning, The Internet of- Things and Big Data", IEEE 26th International Symposium on Industrial Electronics (ISIE), 2017

[21] Vinod Mahor, Romil Rawat, Sachin Chirgaiya, Rabindra Nath Shaw, Ankush Ghosh (2021) "Sentiment Analysis at Online Social Network for Cyber-Malicious Post Reviews Using Machine Learning

Techniques." In: Bansal J.C., Paprzycki M., Bianchini M., Das S. (eds) Computationally Intelligent Systems and their Applications. Studies in Computational Intelligence, vol 950. Springer, Singapore. https://doi.org/10.1007/978-981-16-0407-2_9, (2021)

[22] Rawat, Romil and Mahor, Vinod and Kumar, Anil and TELANG, SHRIKANT and Pachlasiya, Kiran and Garg, Bhagwati and Chouhan, Mukesh, Systematic literature Review (SLR) on Social Media and the Digital Transformation of Drug Trafficking on Darkweb (August 12, 2021). AIBM - 2nd International Conference on "Methods and Applications of Artificial Intelligence and Machine Learning In Heterogeneous Brains" [September 4-6, 2021], Available at SSRN: https://ssrn.com/abstract=3903797 or http://dx.doi.org/10.2139/ssrn.3903797, 2021

[23] Vinod Mahor, Anand Singh Rajawat, Romil Rawat, Rabindra Nath Shaw, Ankush Ghosh "Suspicious Big Text Data Analysis for Prediction—On Darkweb User Activity Using Computational Intelligence Model." In: Mekhilef S., Favorskaya M., Pandey R.K., Shaw R.N. (eds) Innovations in Electrical and Electronic Engineering. Lecture Notes in Electrical Engineering, vol 756. Springer, Singapore. https://doi.org/10.1007/978-981-16-0749-3_58, (2021)

[24] Rawat, R., Rajawat, A. S., Mahor, V., Shaw, R. N., &Ghosh, A. (2021). Surveillance Robot in Cyber Intelligence for Vulnerability Detection. In Machine Learning for Robotics Applications (pp. 107-123). Springer, Singapore.

[25] R. Grossinger, 2018, "5 top benefits & challenges of using Blockchain in healthcare industry," https://www.mtbc.com/learningcenter/Blockchain-technology-benefits-challenges/, 2018.

[26] C. Blenkinsop, "Blockchain's scaling problem, explained," https://cointelegraph.com/explained/Blockchains-scaling-problemexplained, 2018.

[27] Z. Zheng, S. Xie, H. Dai, X. Chen, and H. Wang, "An overview of Blockchain technology: architecture, consensus, and future trends," IEEE International Congress on Big Data, pp. 557–564, 2017.

[28] C. Fernandez-Gago, F. Moyano, and J. Lopez, "Modelling trust dynamics in the internet of things," Information Science, vol. 396, pp. 72–82, 2017.

[29] C. Liu, C. Yang, X. Zhang, and J. Chen, "External integrity verification for outsourced big data in cloud and IoT: a big picture," Future Generation Computer Systems, vol. 49, pp. 58–67, 2015.

[30] J. Yli-Huumo, K. Deokyoon, S. Choi, S. Park, and K. Smolander, "Where is current research on Blockchain technology?—asystematic review," PloS, vol. 11, no. 10, pp. 1–27, 2016.

[31] Berner, C., Brockman, G., Chan, B., Cheung, V., Dębiak, P., Dennison, C., … Jozefowicz, ´ R. (2019). Dota 2 with large scale deep reinforcement learning. arXiv preprint arXiv:1912.06680.

[32] LeCun, Y., Bengio, Y., & Hinton, G. (2015). Deep learning. Nature, 521(7553), 436–444.

[33] Ye, D., Liu, Z., Sun, M., Shi, B., Zhao, P., Wu, H., … Chen, Q. (2019). Mastering complex control in MOBA games with deep reinforcement learning. arXiv preprint arXiv: 1912.09729.

6

Deep Learning Approaches for the Classification of IoT-based Hyperspectral Images

Satyajit Swain* and Anasua Banerjee

School of Computer Engineering, KIIT Deemed to be University, India
Email: swain.satyajit2011@gmail.com; anasua123.banerjee@gmail.com

Abstract

The advancement of unique and efficient knowledge-based expert systems has been significantly influenced by the progress in the field of computer-aided learning and testing, which have demonstrated promising results in a wide range of practical applications. Deep learning techniques, in particular, have been used extensively to detect, predict, and classify patterns in remotely sensed urban land cover areas. Machine learning is already lending a hand in monitoring and tracking changes undergoing on the Earth's surface. However, deep learning enables a higher level of abstraction and provides better extraction of spectral and spatial features in satellite-based imaging analysis. This chapter primarily deals with the application of Internet of Things and deep learning in hyperspectral imaging analysis. For this purpose, some well-known publicly available datasets are used for the classification of different classes present in these datasets. Four main deep neural networks, especially convolution neural network, recurrent neural network (long short-term memory and gated recurrent unit), auto-encoders, and generative adversarial network have been used for the experimentation purpose. A comparative analysis of these classification techniques used for finding the accuracy is made. In the end, certain challenges in deep learning are analyzed, along with some of the emerging future research axes.

6.1 Introduction

In the domain of computing technology, *urban computing* acts as an interdisciplinary field with studies and applications in wireless sensors and networks, big data with computational information, and human–computer interaction of densely populated areas. Many of the ideas established by ubiquitous computing are utilized in urban computing, in which a collection of devices are used to collect data about the urban environment to enhance the quality of life for those who live in cities. The process of acquisition of information about a phenomenon without explicitly making physical contact with the Earth's surface is termed *remote sensing*. The use of satellite or aircraft-based sensor technologies is an integral part of remote sensing to detect and classify objects present on the surface of the Earth, both actively and passively. The diversity of sensors, inputs, and human interaction involved in urban computing sets it apart from typical remote sensing networks. Thus, it brings together inconspicuous and ubiquitous sensor technologies, powerful data management and analytical models, and innovative visualization approaches to offer feasible solutions that enhance the urban environment [1].

The computer visualization field has grown tremendously in recent years as a result of several deep learning (DL) approaches. This was mostly due to two factors. The first is the availability of large image databases containing labeled images and the second is the computing hardware that enabled faster computations [2]. One such popular area of interest in remote sensing is the concept of *hyperspectral imaging* (HSI) which is a novel analytical technique based on spectroscopy and visible imaging. *Spectroscopy* examines the behavior of light sources in a target and recognizes objects based on varied spectral characteristics. It describes how much light is emitted, reflected, or transmitted by a particular item or target [3]. While the human eye lies in the visible spectrum with three color receptors or channels, i.e., red, green, and blue (or RGB), multi-spectral imaging consists of 4–20 channels, only providing discrete and discontinuous portions of the spectral range. HSI, on the other hand, collects hundreds of images at variable wavelengths for the same spatial region creating a hypercube-like structure with a large number of continuous spectral bands. It, thus, gives a complete spectrum for each pixel, as shown in Figure 6.1. Each color denotes different properties of objects present on the Earth's surface. For example, a green color signifies healthy vegetation, whereas a pale color signifies unhealthy, pest infected, and poorly developed vegetation.

The advanced HSI data fields encompass a broad range of processing methods that may successfully extract information from the hyperspectral

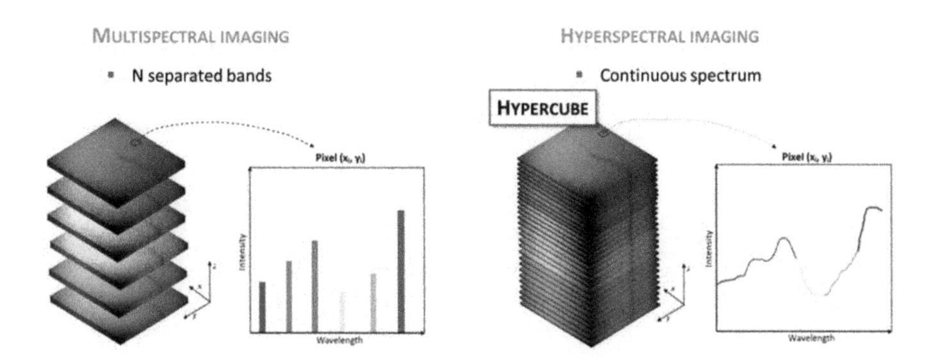

Figure 6.1 Multi-spectral vs. hyperspectral imaging.

cube. Spectral unmixing, resolution enhancement, image restoration and denoising, anomaly detection, dimensionality reduction (DR), and data classification are some of the most used approaches [4]. This chapter mainly focuses on DR and data classification in hyperspectral images. A hyperspectral image consists of a large number of spectral bands (or channels) with several classes of objects present on the Earth's surface. Thus, DR is used as a pre-processing step in the process of classification and detection of objects in these images. Using supervised learning of totally spectrum features within a totally linked architecture, traditional machine learning (ML) methods for HSI classification demonstrate limited performance. DL, on the other hand, provides a wide range of models, including several layers, attributes in both the spatial and spectral domains, and several learning algorithms. The capability to execute a greater number of features makes DL quite robust when handling a huge volume of data, especially in the case of multi-class classification. The main building blocks of a typical DL model comprises convolutional, activation, pooling (or downsampling), and fully connected layers. In this chapter, five different state-of-the-art DL models are used and compared for HSI classification, grouped under convolution neural network (CNN), recurrent neural network (RNN), and compressed network models. All these concepts and models are discussed briefly later in the further sections of this chapter.

The remainder of the chapter is organized as follows. Section 6.2 discusses the use of Internet of Things (IoT) in remote sensing and HSI. In Section 6.3, the prerequisites required for a better understanding of the work are described. The different DL methodologies are presented in Section 6.4, along with the experimental setup in Section 6.5. In Section 6.6, the results obtained are discussed and analyzed. Finally, the chapter concludes in Section 6.7 with a discussion on some future research axes and challenges.

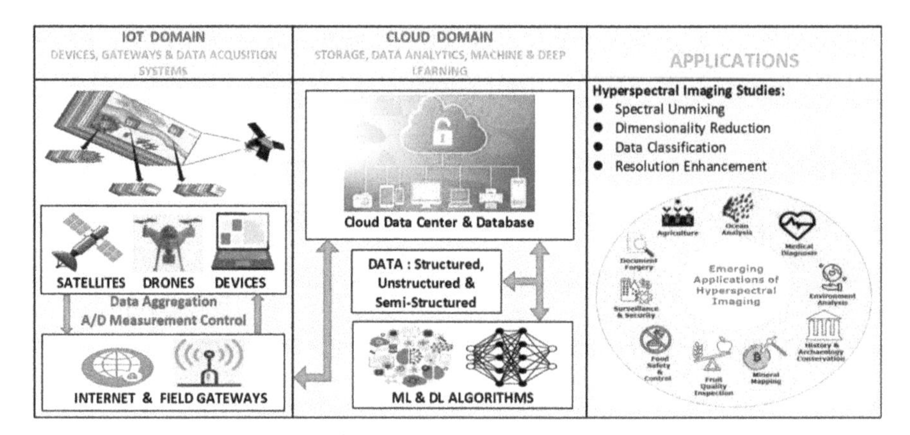

Figure 6.2 IoT in remote sensing and HSI.

6.2 IoT in Remote Sensing and HSI Analysis

IoT being a network of interrelated, internet-connected objects acts as a prominent tool for remote sensing and HSI analysis, which helps to collect and transfer data over a wireless network without human interference. Sensor remote sensing has recently become a major source of information for mapping natural and man-made land coverings. The IoT is created by a communication network in which sensors and actuators integrate into the environment around us and information is transmitted across platforms [5]. Figure 6.2 shows the process of HSI analysis starting from image acquisition to application deployment. The first stage comprises the IoT domain where drones, UAVs, satellites, and devices are used to capture land cover images and scenes from the Earth's surface. These devices aggregate huge volumes of data and communicate with cloud technology, commonly termed edge computing, with the help of different internet protocols and field gateways.

Since hyperspectral images are scenes with large dimensions, they require adequate storage systems to be processed. In the cloud domain, data from the IoT devices are stored in cloud data centers and databases in the form of structured, unstructured, and semi-structured data. Different data analytical techniques including ML and DL models are used to process the huge volume of data. After that, the remotely sensed hyperspectral data is used in studies like spectral unmixing, DR, data classification, etc., and different fields such as healthcare, agriculture, mineral mapping, surveillance, etc. Table 6.1 shows some of the popular hyperspectral sensors used for capturing land cover scenes.

Table 6.1 Some prominently used HSI sensors.

Type of sensor	Sensor name	Frequency	Bands (channels)	GSD	Width
Satellite	PRISMA	0.4–2.5	237	30	≤12
	DESIS	0.4–1	180	30	3.3
	SHALOM	0.4–2.5	241	10	10
	EnMAP	0.42–2.4	228	30	5.25–12.5
	HYPERION	0.4–2.5	220	30	10
Airborne	ROSIS	0.43–0.86	115	1.3	4
	CASI	0.36–1.05	144	2.5	2.4
	AVIRIS	0.36–2.45	224	20	10
	HYMAP	0.4–2.5	126	5	15
	PRISM	0.35–1.05	248	2.5	3.5
	HYDICE	0.4–2.5	210	1–7	10.2

6.2.1 Applications of IoT-based remote sensing and HSI

Remote sensing is one of the fundamental enabling technologies for the IoT, in which nearly any possible item may be endowed with unique identifiers and the capacity to autonomously send data across a network. Remotely sensed IoT-based HSI is enabled in a broad range of applications because of its standoff, label-free, and non-destructive ability in analyzing the elements of matter. Similar to a fingerprint, each substance has a unique spectral signature that may be used to identify it. Thus, HSI is utilized in areas such as molecular biology, mineralogy, astronomy, geology, agriculture, physics, food processing, biomedical imaging, cultural heritage, environment, and surveillance. Also, the use of global positioning systems (GPS) and unmanned aerial vehicles (UAVs) have fostered applications in geographic information systems (GIS) such as video processing and object tracking.

Crop productivity and health under environmental stress, linked illness, crop volatility, soil erosion phases, and agricultural precision are all tracked using it in agriculture. Hyperspectral photography is being tested in the detection of macular edema and retinopathy in the eye care sector, to avoid any harm to the eye. In the food processing business, HSI paired with intelligent software enables automated sorters to identify and eradicate flaws and foreign items that are undetectable by traditional cameras and laser sorters. The geological sample, as well as soil composition analysis, may be easily mapped using HSI for practically all minerals of economic significance. Calcite, feldspar, garnet, olivine, and silica are among the minerals that may be identified from aerial photographs. For numerous monitoring reasons, hyperspectral monitoring and scanning technologies are used. Because military entities use precautions to prevent airborne monitoring, HSI is quite useful for military surveillance. Water quality study, precipitation, and the detection of sea ice in water and marine resource management have all been the topic of several studies. Soldiers are exposed to a variety of chemical threats, most of which are undetectable but may be detected by HSI technology. In forest and environmental management, HSI is used to assess forest health and condition as well as to detect protruding species and forest plantation infestation. It is also used to monitor pollution from coal- and oil-fired power stations, hazardous waste incinerators, cement mills, and other industrial sources.

6.2.2 Challenges in IoT-based remote sensing and HSI

Despite several wide-ranging applications, IoT-based remote sensing faces certain challenges in terms of processing and handling of data. Huge volumes of data, complicated data formats, various map projections, GIS applications, communication capabilities, and processing time are all critical challenges that must be addressed before remote sensing data can be widely used. Preprocessing and validation are the two most arduous facets of remote sensing technology [6]. For site-specific studies, although their resolutions differ, a variety of remote sensing data packages are available. Forest fire assessment, for example, necessitates high geographical and temporal resolution data. Researchers in many disciplines have used a wide variety of models based on remote sensing data. All mathematical models are not compatible with all sensors, nor do they operate in all geographical areas. As a result, *in situ* data must be used to validate the model. Handling huge amounts of data in complex forms with complicated processing is a big challenge that must be addressed. The precise positioning of physical items via IoT is critical for a variety of location-based services and applications [7]. For traceability, precision, and dependability, smart sensors that are utilized for monitoring, measurement, and control require accurate calibration techniques and standards. Both physical and digital securities also play a crucial challenge where sensors and UAVs are used in open fields to tackle the challenging environment and protect the data.

In the case of HSI, cost and complexity are the two biggest drawbacks. For hyperspectral data analysis, fast processors, sensitive detectors, and enormous data storage capacity are required to handle large uncompressed data with huge dimensionality. Programming hyperspectral sensors and satellites to filter through data on their own and transmit just the most essential information is also a challenge because transmitting and storing that much data may be challenging and expensive [8]. Hyperspectral images can lead to significant variability in target and background spectra that can make the classification task difficult. The intra-class variability in HSI data is significant, resulting in uncontrolled variations in spectrometer-captured reflectance which is usually due to variations in atmospheric situations. The pixels in hyperspectral data enfold wider geographic areas on the Earth's surface than images with low or medium spatial resolution. As a result, they appear to exhibit mixed spectral signatures in boundary zones, with more inter-class similarity. Thus, it leads to a pixel pureness challenge where two samples of the same class appear to be different, and two samples of different classes appear to be same. Also, the spectrometer's instrumental noise can degrade the data gathering process, distort the spectral channels to variable degrees, or exhibit many channels useless due to calibration or saturation issues.

6.3 Preliminaries

Hyperspectral images can be represented as H, $H \in R^{L \times B \times D}$, where $L \times B$ denotes the spatial dimension and D denotes the number of large spectral bands. It can be represented in the form of a 3D hyperspectral cube. In this section, various DR techniques, DL models, and activation function used are explained for prior understanding.

6.3.1 Dimensionality reduction

The dimension of a dataset demotes the total number of variables or features present in it. If many numbers of features and variables are present in the hyperspectral dataset, the model tends to overfit and cannot generalize well, thus making the classification process difficult. To overcome this problem of *curse of dimensionality*, many linear and non-linear DR techniques are applied to the raw input data. Two different DR techniques have been used here.

6.3.1.1 Principal component analysis (PCA)

PCA is a linear unsupervised feature selection technique, which, if detects two features in a dataset correlating to each other, merges both the features into a new single feature [9]. As a result, the redundant information gets removed from the dataset. First, the mean values of the given M number of samples are calculated, and the mean adjusted matrix values are gathered. In the next stage, the co-variance matrix along with the eigenvalues and corresponding eigenvectors are computed. Finally, the basic vector values are determined with the help of a scatter matrix P_a of the eigenvectors, as given in the following equation:

$$P_a = \sum_{i=1}^{A} (X_i - \sigma).(X_i - \sigma)^a \tag{6.1}$$

where σ represents the mean of all the data and X_i is the *i*th image with its column in a vector.

6.3.1.2 Kernel PCA

It is an extension of PCA, applying kernel techniques in the field of multivariate statistics [10]. Being a non-linear DR technique, it works better in both linear as well as non-linear input data, unlike PCA, which works well only on linear input data. The mathematical equation of KPCA is shown in the following equation:

$$\alpha.\delta = K.\delta \tag{6.2}$$

where K is the kernel function, α is the eigenvalue, and δ is the eigenvector of α.

6.3.2 Deep learning models

In this chapter, five different types of DL models are used and compared for the classification purpose: gated recurrent unit (GRU), 3D CNN, long short-term memory (LSTM), auto-encoders (AE), and generative adversarial networks (GAN).

6.3.2.1 Gated recurrent unit

GRU is a type of RNN model which follows a gating mechanism and works better for smaller datasets. It is composed of two gates, namely reset gate and update gate, as shown in Figure 6.3. It does not keep track of the internal cell state and lacks an output gate, unlike LSTM [11]. The function of the update

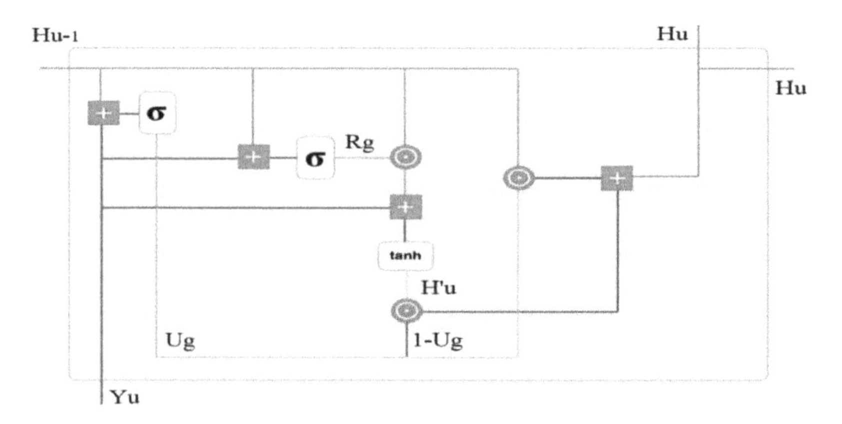

Figure 6.3 GRU internal architecture.

gate is to decide which information is to be passed along to the future. The update gate U_g for time step u is given in the following equation:

$$U_g = \sigma(W_U Y_u + O_U H_{u-1}) \tag{6.3}$$

where σ is the sigmoid function and input Y_u is multiplied by its weight W_U. Similarly, H_{u-1} holding information for previous u–1 units is multiplied by its weight O_U. The work of the reset gate is to determine what amount of information received from the preceding state is to be discarded. The reset gate R_g for time step u is given in the following equation:

$$R_g = \sigma(W_R Y_u + O_R H_{u-1}) \tag{6.4}$$

Finally, summation operation takes place and it is passed through tanh activation function.

6.3.2.2 Long short-term memory

LSTM is also a type of RNN model following the gating mechanism [12]. Unlike GRU, LSTM consists of three gates, namely input gate, forget gate, and output gate, as shown in Figure 6.4. The input gate determines which values need to be updated after receiving values from the previous state and current input. Forget gate F_g helps to take the decision on which pieces of information are to be deleted by multiplying the previous cell state so that unnecessary information is not carried forward to the new state, as given in the following equation:

$$F_g = \delta\left(W_F \left(H_{(t-1)}, Y_t\right) + b_F\right) \tag{6.5}$$

where $H_{(t-1)}$ is the hidden state input from the previous cell and Y_t is the current timestamp input. The new information gathered from the present input is

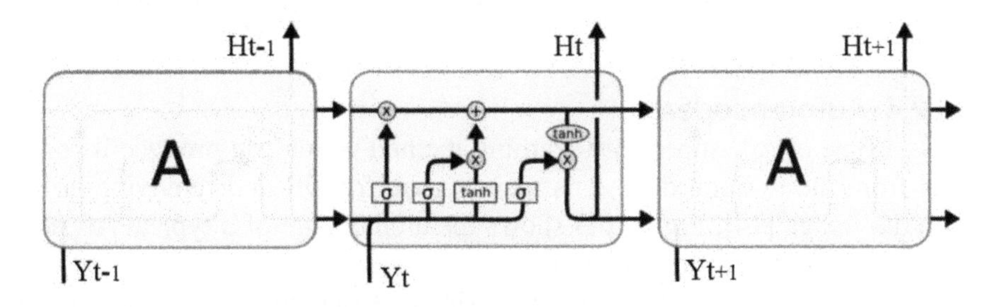

Figure 6.4 LSTM internal architecture.

added to the present cell. Here, a sigmoid layer δ determines which values are to be updated or not, and a tanh layer generates a vector for new candidates to add to the current cell state, as given in the following equations:

$$I_g = \delta\left(W_I \left(H_{(t-1)}, Y_t\right) + b_I\right) \tag{6.6}$$

$$T_g = \tanh\left(W_T \left(H_{(t-1)}, Y_t\right) + b_T\right) \tag{6.7}$$

where T_g is the new candidate value. Finally, the output O_g and input for the next layer H_t are decided based on the cell state, as given in the following equations:

$$O_g = \delta\left(W_O \left(H_{(t-1)}, Y_t\right) + b_O\right) \tag{6.8}$$

$$H_t = O_g * \tanh\left(T_g\right) \tag{6.9}$$

6.3.2.3 3D CNN

3D CNN has the ability to learn automatically various features from the input dataset with the help of various layers like convolution, pooling, dense, flatten, and fully connected layer. 3D CNN as compared to 2D CNN is capable of extracting both spectral and spatial features but is more complex [13]. The equation for 3D CNN is given as follows:

$$A_{p,q,r}^{u,v,w} = \gamma\left(b_{p,q} + \sum_{\lambda=1}^{f_{p-1}} \sum_{\varphi=-\alpha}^{\alpha} \sum_{\pi=-\beta}^{\beta} \sum_{\theta=-\psi}^{\psi} W_{lp,q,r}^{\theta,\pi,\beta} A_{p-1,\gamma}^{u+\theta,v+\pi,c+\gamma}\right) \tag{6.10}$$

where Υ denotes the activation function, b represents the bias function for qth feature map of rth layer, Ψ and β are the height and width of the kernel, u, v, and w represent the spatial position of the pth layer for the qth feature map, defined by .

6.3.2.4 Auto-encoders

AE is a type of self-supervised learning method which can produce its own label from the training data. It is mainly used for DR and removing noise from the image [14]. Figure 6.5 shows the architecture of a typical AE network. It consists of four main parts.

(a) Encoder: It encodes the input dimensions into a compressed and reduced form.

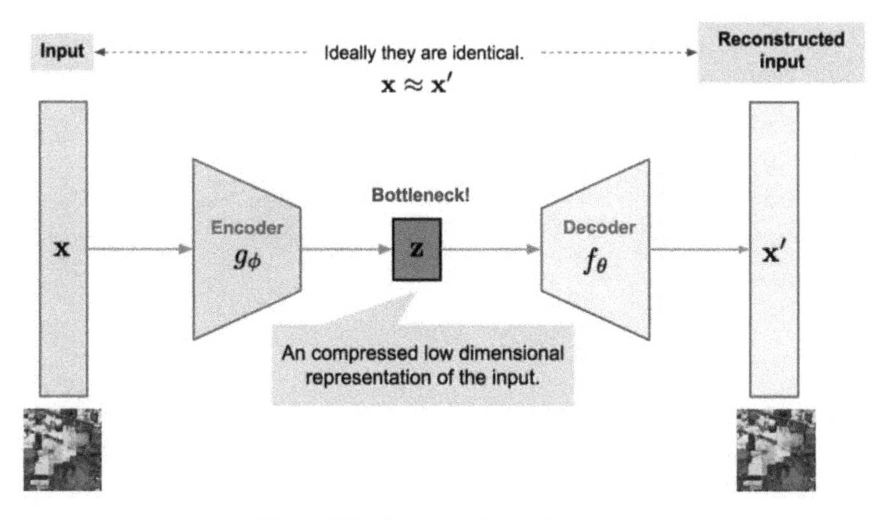

Figure 6.5 Auto-encoder architecture.

(b) Bottleneck: It contains the compressed information of the input data in the lowest possible dimension.

(c) Decoder: It decodes the encoded information by reconstructing the data as similar as possible to the actual input.

(d) Reconstruction loss: It estimates the performance of the decoder.

In an AE model, the encoder represented as g_Φ maps original input x into a latent space z present at the bottleneck, given as $g_\Phi: x \rightarrow z$. Similarly, the decoder represented as f_θ maps the latent space to the output x', given as $f_\theta: z \rightarrow x'$, where $x \approx x'$. The encoded and decoded representation for the input is given in eqn (6.11) and (6.12), respectively.

$$z = \alpha(W.x + b) \tag{6.11}$$

$$x' = \alpha'(W'.z + b') \tag{6.12}$$

where *W, W′* denotes the weights, and *b, b′* denotes the bias. The reconstruction loss representation, denoted as *L*, is given in the following equation:

$$L(x,x') = \|x - x'\|^2 = \left\|x - \alpha'\left(W'\left(\alpha(W.x+b)\right)+b'\right)\right\|^2 \tag{6.13}$$

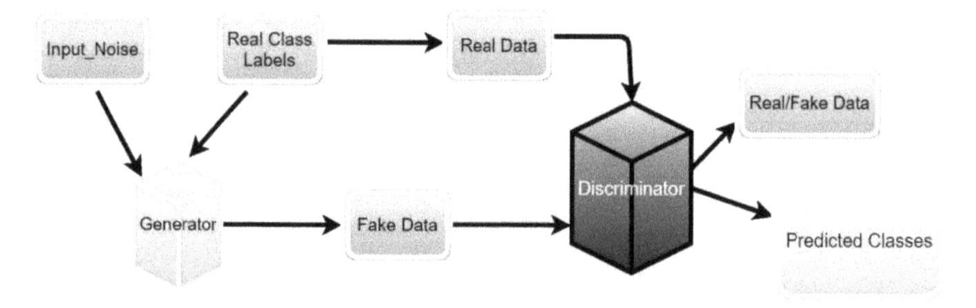

Figure 6.6 GAN architecture.

6.3.2.5 Generative adversarial network

GAN is a type of supervised model which comprises two neural networks, namely generator and discriminator, as shown in Figure 6.6. A generator network is mainly used to generate fake samples with the help of some random noise and actual class labels. However, a discriminator is used to distinguish between fake and real samples. At the time of training, the GAN generator tries to make fools by generating more realistic fake samples, whereas the discriminator seeks to avoid being fooled by the generator [15]. To train the GAN model, a minimax game algorithm is performed, where both the generator and discriminator networks are trained simultaneously.

The discriminator part is also used for classification since it maps features to labels. The equations for the GAN model are given as follows:

$$I_H = N\lfloor \log \mathrm{J}(H = \mathrm{real}|T_{\mathrm{real}})\rfloor + N\lfloor \log \mathrm{J}(H = \mathrm{fake}\,|\,T_{\mathrm{fake}})\rfloor \quad (6.14)$$

$$I_O = N\lfloor \log \mathrm{J}(O = O|T_{\mathrm{real}})\rfloor + N\lfloor \log \mathrm{J}(O = O\,|\,T_{\mathrm{fake}})\rfloor \quad (6.15)$$

where I_H denotes the log-likelihood right source of input data and I_O denotes the log-likelihood right of class labels.

6.3.3 Activation function

Activation functions are mostly used to determine which outputs should be fired, as well as to normalize outputs that range from 1 to 0 or –1 to 0. Unlike linear activation functions, the non-linear ones can train the model using back-propagation functionality. They are commonly used in complex models, such as the classification of high-dimensional image and video datasets. In this chapter, a smooth non-linear activation known as Mish is used.

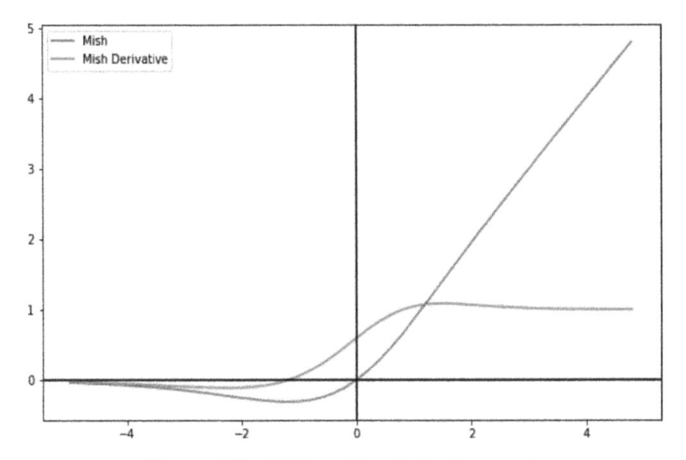

Figure 6.7 Mish and its derivative graph.

The Mish function is basically *c* times of combination of tanh and softplus function, where *c* is the input, and the output ranges from –1 to 1. It is a zero-centered function because it has an equal mass on both sides of the zero-axis. As a result, it can solve the sigmoid activation function problem, making it preferable to the Swish activation function. The Mish function is given in (6.16), with its graph in Figure 6.7.

$$f(c) = c * \tanh\left(\ln\left(1 + e^{c}\right)\right) \tag{6.16}$$

6.4 Methodology

In this section, the architectures of different DL models used are discussed. A hyperspectral input image with large spectral dimensions is passed on to a DR technique, where the large number of bands present in the image is reduced prior to classification. The image obtained after reducing the dimensions is passed on to the DL model for classification with several layers, including convolution, pooling, dense, and fully connected layers. The black box, as shown in Figure 6.8, represents the different DL models used in this chapter for training purposes.

6.4.1 Gated recurrent unit

Since a GRU model works better with 1D input, the 3D input image with dimensions (W_S, W_S, K, 1) is first reshaped to 1D before passing it to GRU layers, where W_S denotes the window size and K denotes the reduced spectral

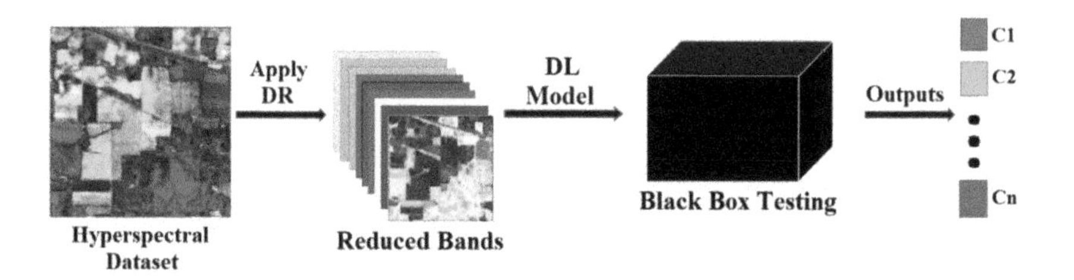

Figure 6.8 HSI classification process.

bands. In this model, three GRU layers with units 50, 100, and 150 are used sequentially, respectively. The activation function used in these layers is taken as Mish. Followed by that, two dense layers with units 128 and 64, and dropout layers with size 0.4 are used. Finally, a dense layer with a size equal to the number of classes present in the hyperspectral image is used, with activation function as Softmax.

6.4.2 Long short-term memory

A bidirectional LSTM architecture is used, which can learn the input sequence in both forward and backward directions and can also concatenate both the interpretations. It does so by wrapping the LSTM layers in a wrapper layer known as *bidirectional*. Similar to the functioning of GRU, three LSTM layers with units 50, 100, and 150 are used with activation function as Mish. Followed by that, two dense layers with units 128 and 64 and dropout layers with size 0.4 are used. Finally, a dense layer with a size equal to the number of classes present in the hyperspectral image is used, with activation function as Softmax.

6.4.3 3D CNN

Since a 3D CNN requires a 3D input, the input image (W_S, W_S, K, 1) is passed on to a 3D convolution layer with kernel size taken as ($7 \times 7 \times 5$) with 16 number of filters. Followed by that, another convolution layer with kernel size ($5 \times 5 \times 3$) with 32 filters is used. Similarly, another convolution layer with kernel size ($3 \times 3 \times 3$) with 64 filters is used. The activation in all the three convolution layers is taken as Mish. The output from the convolution layer is passed on to a 3D max pooling layer with size ($3 \times 3 \times 3$). After that, a flatten layer is used to convert the 3D output to 1D. Followed by that, two dense layers with units 256 and 128, and dropout layers with size 0.4 are

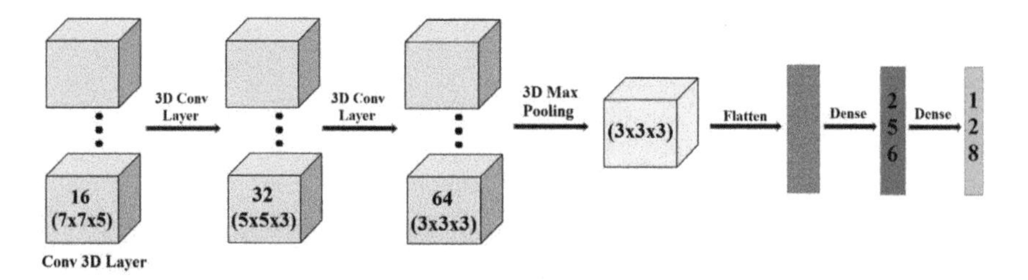

Figure 6.9 3D CNN layers.

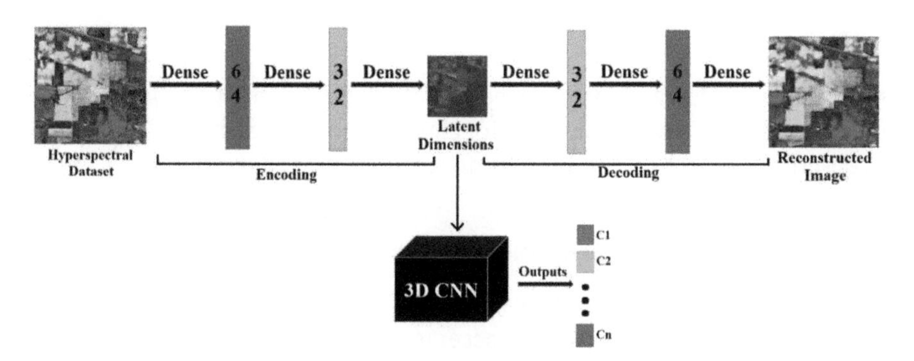

Figure 6.10 AE architecture.

used. Finally, a dense layer with a size equal to the number of classes present in the hyperspectral image is used, with activation function as Softmax. The black box architecture for 3D CNN is shown in Figure 6.9.

6.4.4 Auto-encoder

A convolution AE is used in this chapter which works better for hyperspectral images. Since AE follows an encoding–decoding process, the architecture is divided into two parts. The first part is the encoding–decoding process where the model is trained to minimize the loss function. Since an AE is used for both DR and classification purposes, no other DR technique is applied here for reducing the large dimensionality of hyperspectral images. The encoder network first converts the 3D input into a latent reduced dimension, with three sequential dense layers with units 64, 32, and 15, respectively. Followed by that, a decoder network then reconstructs the latent representation into the original image with dimensions same as that of the input, working in reverse chronological order. The architecture for the AE model is shown in Figure 6.10. The second part is the classification task where the decoder part from the trained

Table 6.2 3D GAN layers.

Network	Layers	BN	Stride
Generator	$4 \times 4 \times 256$	Yes	1
	$4 \times 4 \times 128$	Yes	2
	$4 \times 4 \times 64$	Yes	2
	$4 \times 4 \times K$	No	2
Discriminator	$4 \times 4 \times 64$	No	2
	$4 \times 4 \times 128$	Yes	2
	$4 \times 4 \times 256$	Yes	2
	$4 \times 4 \times 32$	No	1
	$32 \times C$	No	

model is discarded and the encoded representation with latent dimensions is thus passed on to any DL classifier for classification. Here, the 3D CNN model as described in Section 6.4.3 is used for classification purpose.

6.4.5 Generative adversarial network

A 3D GAN model is used in this chapter for HSI classification. In the first layer of the GAN, the generator accepts a uniform noise distribution n as input which is basically a matrix multiplication. It then transforms its shape into a tensor of same size as that of real data with K reduced components, which is used at the start of the generator stack. The discriminator accepts the real class labels and the fake samples generated by the generator network to classify the C class labels using Softmax classifier. Table 6.2 shows the 3D convolution layers used in both the generator as well as discriminator networks. Here, strided convolutions are used in both the networks which act as a replacement for pooling layers. Since GANs are complex models with slow training, batch normalization has been used in some layers to accelerate the training process by reducing the generalization error.

6.5 Experimental Setup

6.5.1 Hyperspectral datasets

Several public datasets obtained with hyperspectral sensors have been made accessible so far. Table 6.3 entails the different hyperspectral datasets commonly used for different HSI applications. Out of those, three well-known datasets, i.e., Indian Pines (IP), Pavia University (PU), and Salinas Valley (SV), are employed in this chapter for experimental purposes [16]. Figure 6.11 shows the total number of class-wise samples for each dataset used.

Table 6.3 Popular HSI datasets.

Dataset	Indian Pines	Pavia University	Salinas Valley	KSC	Houston	Botswana
*SD	145 × 145	610 × 340	512 × 217	614 × 512	349 × 1905	1476 × 256
Bands	200	103	224	176	144	145
Range	0.4–2.5	0.43–0.86	0.4–2.5	0.4–2.5	0.36–1.05	0.4–2.5
*SR	20 m	1.3 m	3.7 m	18 m	2.5 m	30 m
Samples	10,249	42,776	54,129	5211	15,011	3248
Classes	16	9	16	13	15	14
Mode	Aerial	Aerial	Aerial	Aerial	Aerial	Satellite
Sensor	AVIRIS	ROSIS	AVIRIS	AVIRIS	CASI	HYPERION

*SD denotes spatial dimensions and SR denotes spatial resolution.

S. No.	Indian Pines	Samples	Pavia University	Samples	Salinas Valley	Samples
1.	Alfalfa	46	Asphalt	6631	Brocoli_green_weeds_1	2009
2.	Corn-notill	1428	Meadows	18,649	Brocoli_green_weeds_2	3726
3.	Corn-mintill	830	Gravel	2099	Fallow	1976
4.	Corn	237	Trees	3064	Fallow_rough_plow	1394
5.	Grass-pasture	483	Painted metal sheets	1345	Fallow_smooth	2678
6.	Grass-trees	730	Bare Soil	5029	Stubble	3959
7.	Grass-pasture-mowed	28	Bitumen	1330	Celery	3579
8.	Hay-windrowed	478	Self-Blocking Bricks	3682	Grapes_unstrained	11,271
9.	Oats	20	Shadows	947	Soil_vinyard_develop	6203
10.	Soybean-notill	972			Corn_senesced_grren_weeds	3278
11.	Soybean-mintill	2455			Lettuce_romaine_4wk	1068
12.	Soybean-clean	593			Lettuce_romaine_5wk	1927
13.	Wheat	205			Lettuce_romaine_6wk	916
14.	Woods	1265			Lettuce_romaine_7wk	1070
15.	Buildings-grass-Trees	386			Vinyard_untrained	7268
16.	Stone-Steel-Towers	93			Vinyard_treils	1807
Total Samples		10,249		42,776		54,129

Figure 6.11 Class-wise samples for HSI datasets used.

6.5.2 Experimental setup and parameters

All the experiments in this chapter were conducted in Google Colab with Python 3.6 as the programming language, TensorFlow package version of 2.3.0, Keras framework version 2.4.3, RAM size of 25.51 GB, disk size of 68.4 GB, and a GPU hardware accelerator run time.

The experiments were carried out on the three hyperspectral datasets, as mentioned in Section 6.5.1. A standard scaler was used for scaling the datasets as a pre-processing step. The large number of spectral bands in each dataset was reduced to 15 using PCA and kernel PCA techniques. Figures 6.12–6.14 show five hyperspectral bands for each dataset, before and after applying the DR techniques, respectively. Here, each band represents a group of wavelengths with its unique properties based on different colors. In HSI classification, the hyperspectral data cube must be partitioned into compact overlapping 3D neighboring patches with 2D spatial extent, also known as the window size, to adjust the parameters of the kernels for each pixel centered in that layer. Here, the window size was taken as 25 due to limiting RAM size. The training and testing ratios for the datasets were taken as follows: Indian Pines (10%, 90%), Pavia University (5%, 95%), and Salinas Valley (5%, 95%). The activation function used in all the DL models was chosen as Mish and Softmax (for final output layer only). Other parameters include the model optimizer which is the Adam optimizer with a learning rate of 0.001, the loss function being categorical cross-entropy, and metrics being the accuracy. The batch size was taken as 200 and 250 epochs were run with ten simulations for each category, using the proposed models for calculating the overall accuracy (OA), average accuracy (AA), and kappa accuracy (KA). The OA is the ratio of the number of accurately predicted samples to the total number of samples. The AA is the mean accuracy of all the predicted classes.

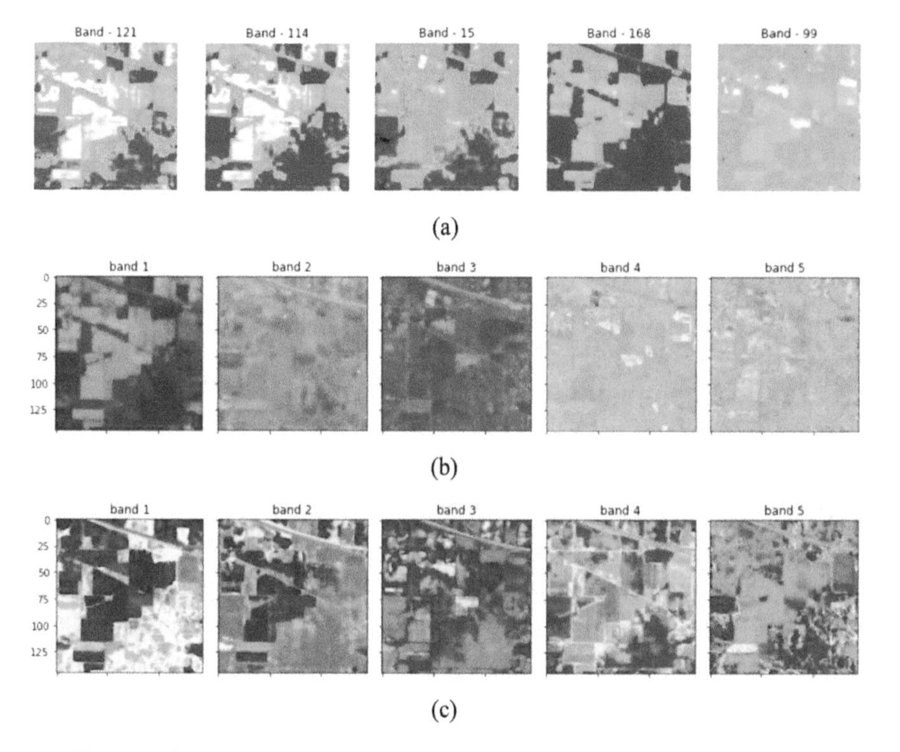

Figure 6.12 IP dataset: (a) before DR, (b) after PCA, and (c) after kernel PCA.

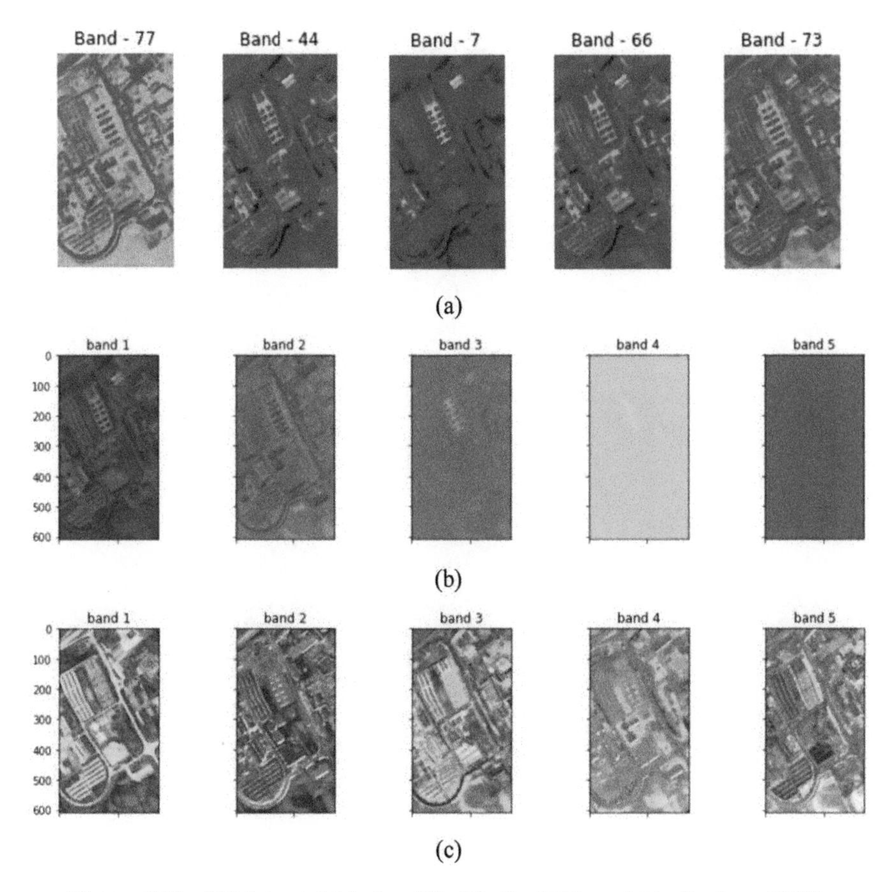

Figure 6.13 PU dataset: (a) before DR, (b) after PCA, and (c) after kernel PCA.

The KA denotes the relation between the predicted image and the ground truth image.

6.6 Results and Discussion

The results obtained after applying all the parameters and metrics for all the three datasets, i.e., IP, PU, and SV, are shown in Tables 6.4–6.6, respectively. All these tables show values obtained in terms of OA, AA, and KA, using the different parameters as discussed in Section 6.5.2. The highest accuracy obtained in all the datasets is marked in red bold. Similarly, the second-best accuracy obtained in all the datasets is marked in green bold. Since the AE model did not use either PCA or kernel PCA, its accuracy values are arranged sequentially as OA, AA, and KA, respectively.

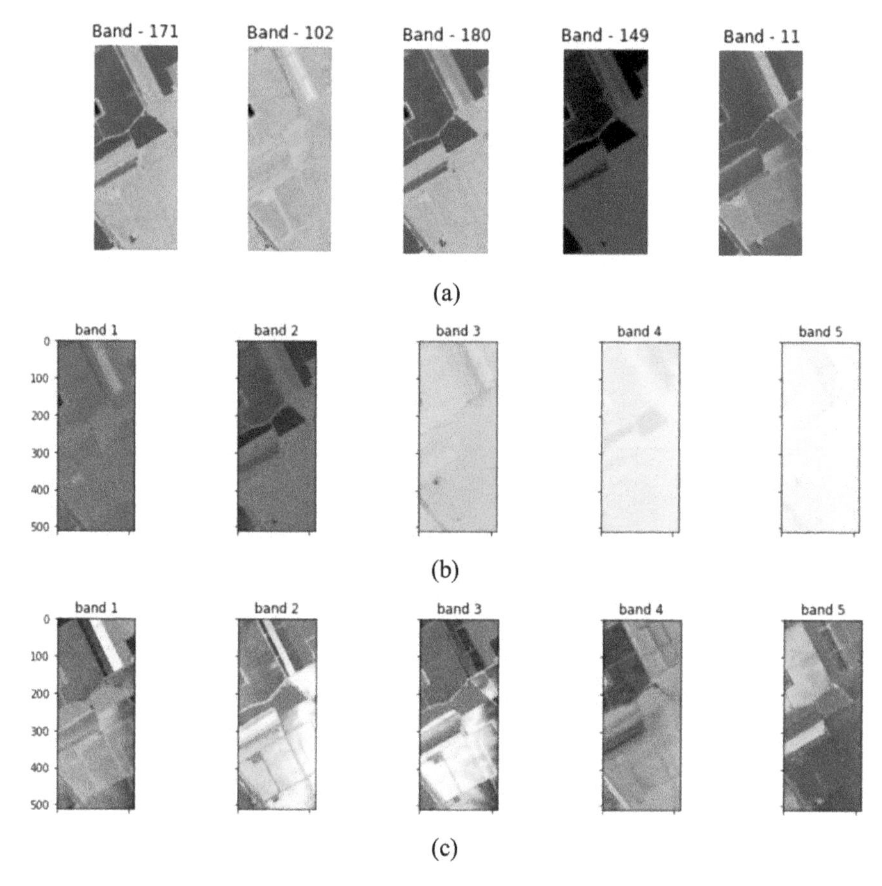

Figure 6.14 SV dataset: (a) before DR, (b) after PCA, and (c) after kernel PCA.

From the above tables, it can be seen that the highest accuracy is obtained using the combination of AE with 3D CNN model for all the datasets. AE makes use of neural networks, such as CNN for the task of representation learning of input data by transforming it, which makes an effective model for HSI classification. It finds different levels of features since it tries to learn a lot of useful information and non-linear transformations, reducing noise and redundancy from the data along with minimizing the loss. 3D CNN, with its powerful ability for edge detection and spatial correlation, is able to perform better than GAN and RNN models since it can extract both spectral and spatial features from hyperspectral data quite efficiently. Thus, the combination of AE with 3D CNN provides quite better accuracy as compared to the simple 3D CNN model. GAN being a very complex neural network model is quite difficult to train. This is because both the generator

Table 6.4 IP dataset accuracy table.

DL model	PCA			Kernel PCA		
	OA	AA	KA	OA	AA	KA
GRU	94.46 ± 0.3	94.42 ± 0.3	94.4 ± 0.3	95.14 ± 0.4	95.11 ± 0.3	95.09 ± 0.4
LSTM	94.67 ± 0.3	94.63 ± 0.4	94.62 ± 0.3	94.98 ± 0.4	94.96 ± 0.4	94.96 ± 0.3
GAN	95.97 ± 0.2	95.95 ± 0.3	95.92 ± 0.3	96 ± 0.1	95.94 ± 0.3	95.91 ± 0.2
3D CNN	98.01 ± 0.0	97.99 ± 0.0	97.94 ± 0.0	**98.29 ± 0.0**	**98.25 ± 0.0**	**98.23 ± 0.0**
AE + 3D CNN	**99.16 ± 0.0**		**99.13 ± 0.0**		**99.11 ± 0.0**	

Table 6.5 PU dataset accuracy table.

DL Model	PCA			Kernel PCA		
	OA	AA	KA	OA	AA	KA
GRU	95.16 ± 0.3	95.11 ± 0.3	95.09 ± 0.3	95.37 ± 0.4	95.35 ± 0.3	95.31 ± 0.4
LSTM	95.03 ± 0.3	94.98 ± 0.4	94.97 ± 0.3	95.43 ± 0.4	95.4 ± 0.4	95.38 ± 0.3
GAN	96.07 ± 0.2	96.02 ± 0.3	96 ± 0.3	96.19 ± 0.1	96.14 ± 0.3	96.13 ± 0.2
3D CNN	98.18 ± 0.0	98.14 ± 0.0	98.14 ± 0.0	**98.55 ± 0.0**	**98.52 ± 0.0**	**98.48 ± 0.0**
AE + 3D CNN	**99.47 ± 0.0**		**99.45 ± 0.0**		**99.41 ± 0.0**	

Table 6.6 SV dataset accuracy table.

DL model	PCA			Kernel PCA		
	OA	AA	KA	OA	AA	KA
GRU	95.21 ± 0.3	95.18 ± 0.3	95.15 ± 0.3	95.55 ± 0.4	95.52 ± 0.3	95.5 ± 0.4
LSTM	95.32 ± 0.3	95.3 ± 0.4	95.27 ± 0.3	95.8 ± 0.4	95.78 ± 0.4	95.77 ± 0.3
GAN	96.3 ± 0.2	96.26 ± 0.3	96.24 ± 0.3	96.32 ± 0.1	96.29 ± 0.3	96.25 ± 0.2
3D CNN	98.59 ± 0.0	98.56 ± 0.0	98.54 ± 0.0	**98.83 ± 0.0**	**98.8 ± 0.0**	**98.77 ± 0.0**
AE + 3D CNN	**99.68 ± 0.0**		**99.66 ± 0.0**		**99.63 ± 0.0**	

and discriminator networks are trained simultaneously in a zero-sum game, which makes the model unstable and fails to converge. RNN models being more suitable for sequence predictions do not perform well in case of HSI classification. Due to their recurrent nature, they are difficult to train and undergo slow computations.

In case of DR technique, kernel PCA being a non-linear technique performs better than PCA. It extracts non-linear features from hyperspectral

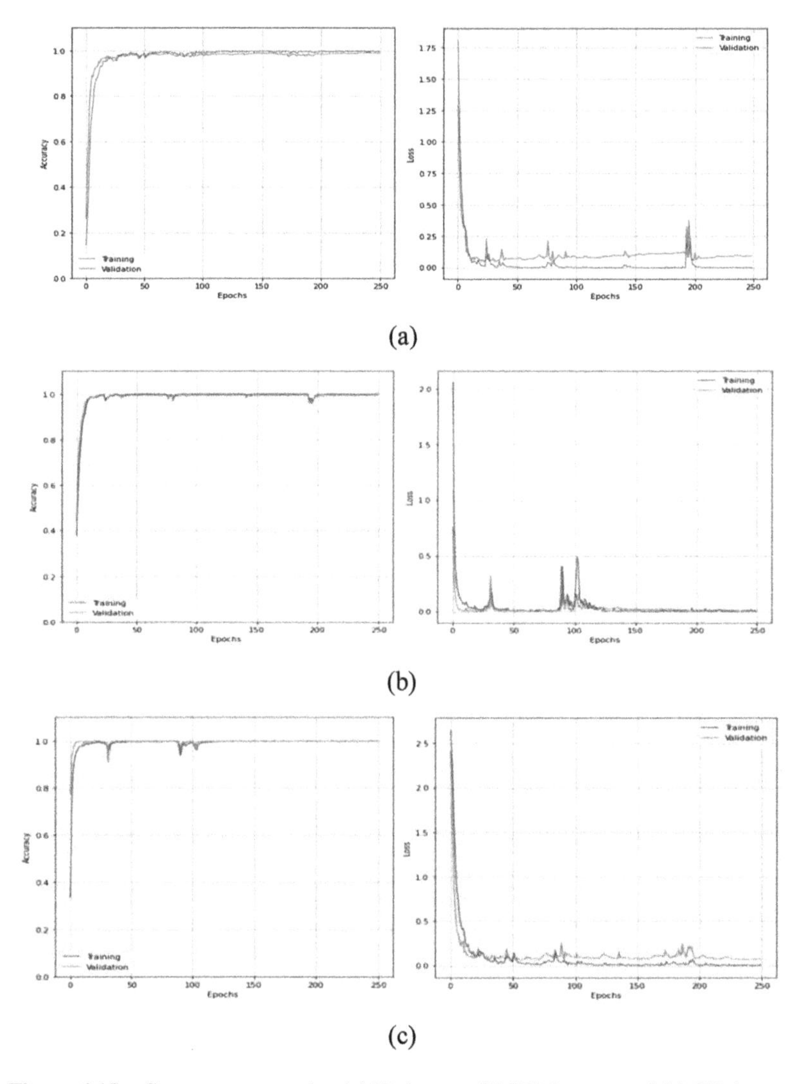

Figure 6.15 Convergence graphs: (a) IP dataset, (b) PU dataset, and (c) SV dataset.

images and enhances classification efficiency by removing non-linearity in the data. Also, it projects the linearly separable data into a high-dimensional feature space, unlike PCA which works well only on linearly separable data. It also tries to capture both local and global changes in the data, unlike PCA which captures only global changes. Figure 6.15 shows the accuracy and loss convergence graph obtained for the three datasets using the best model and DR technique obtained, respectively.

The information about a certain site is captured using a ground truth image. Because of the presence of poor atmospheric conditions and other barriers, information for a hyperspectral image is distorted when acquired from satellite or aerial mode. Thus, the ground truth image information is used as a reference for overcoming issues caused by distorted information. Figures 6.16–6.18 show the ground truth and predicted images for all three datasets, respectively. It can be visualized that the predicted images via AE and 3D CNN models resemble strong similarity to the corresponding ground truth image as compared to other DL models.

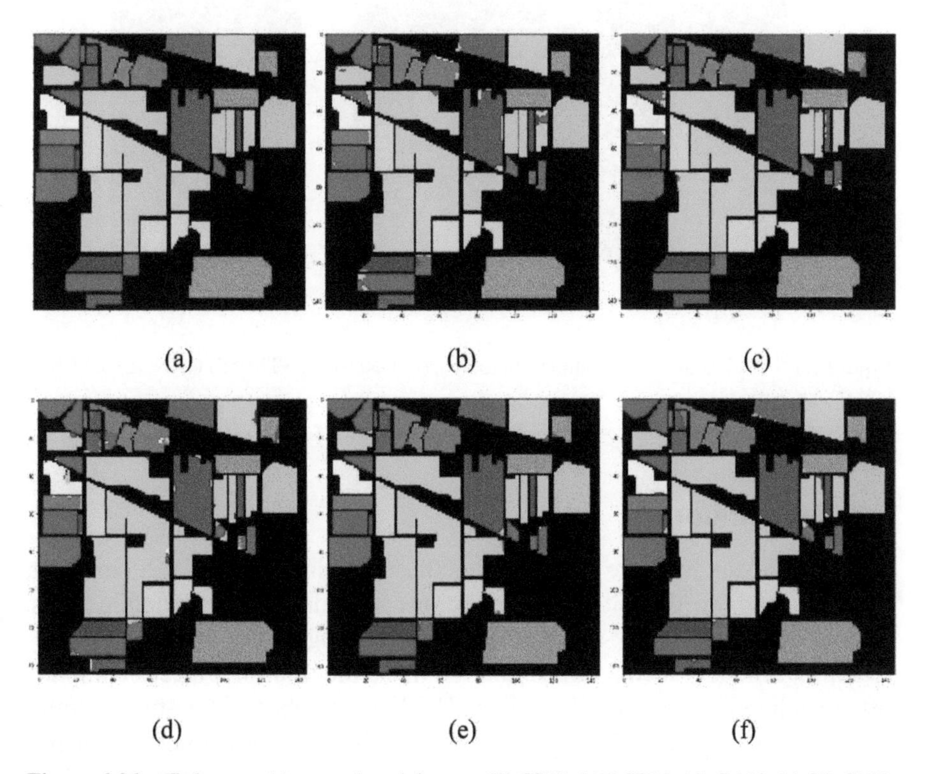

Figure 6.16 IP dataset: (a) ground truth image, (b) GRU, (c) LSTM, (d) GAN, (e) 3D CNN, and (f) AE + 3D CNN.

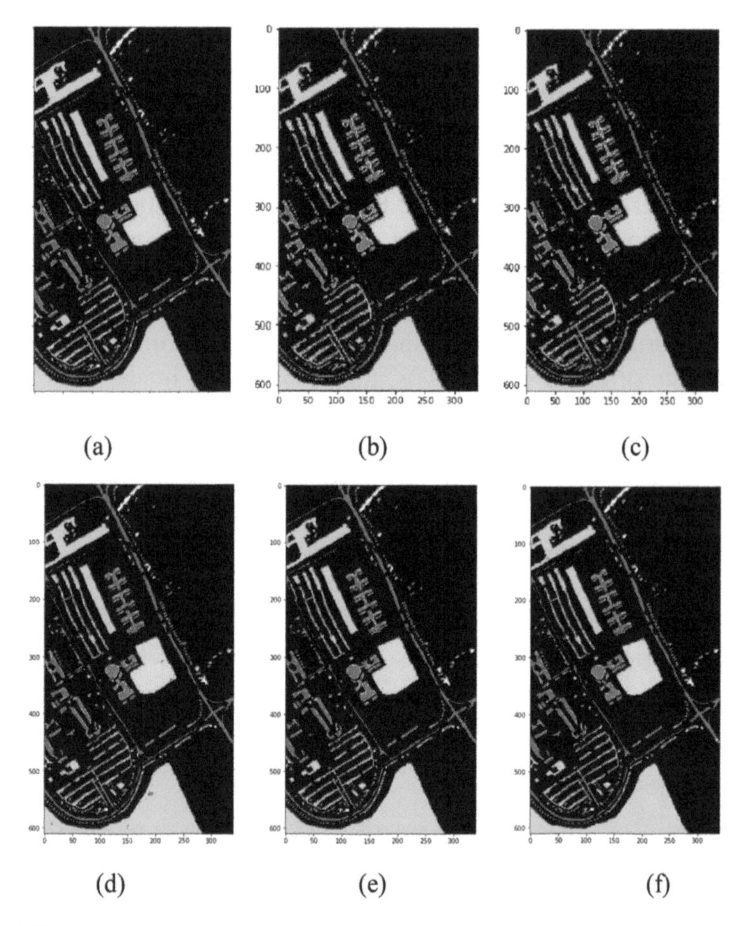

Figure 6.17 PU dataset: (a) ground truth image, (b) GRU, (c) LSTM, (d) GAN, (e) 3D CNN, and (f) AE + 3D CNN.

6.7 Conclusion

HSI poses many intricate challenges as discussed in Section 6.2.2. However, DL also tends to exhibit some limitations in neural network applications. The non-convex and NP-complete nature of tuning and optimizing parameters in DL models such as GAN and AE make the training process complex, often leading to severe local minima. The large number of fine-tuning training parameters procured by supervised models makes it memory intensive, leading to increased computational load. While forward propagation leads to considerable data loss, backward propagation faces issues in the propagation

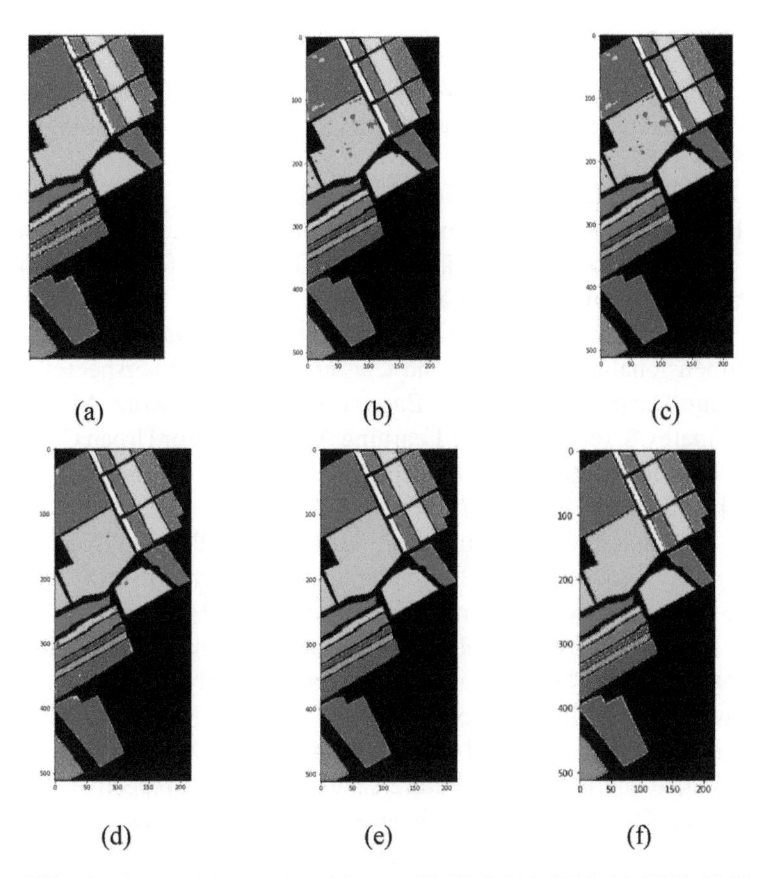

Figure 6.18 SV dataset: (a) ground truth image, (b) GRU, (c) LSTM, (d) GAN, (e) 3D CNN, and (f) AE + 3D CNN.

of gradient and activation signals in deep neural networks, often leading to vanishing gradient problems.

This chapter mainly focused on two studies of HSI, i.e., DR and data classification, where five state-of-the-art DL models were used and compared on the basis of classification accuracy obtained by those models. AE combining with 3D CNN classifier performs quite well combining both the encoding and correlating features of AE and 3D CNN, respectively. Similarly, the combination of 3D CNN with kernel PCA also proves quite effective for HSI classification. In addition, the Mish function also enhances the classification accuracy, generating smooth feature maps for obtaining better-predicted images. However, better band optimization techniques for DR can be used along with improvements in the GAN model as a part of future research studies.

References

[1] Y. Zheng, L. Capra, O. Wolfson, and H. Yang, "Urban Computing", ACM Transactions on Intelligent Systems and Technology, 1–55, ISSN 2157-6904, 2014.

[2] M. A. Ponti, L. S. F. Ribeiro, T. S. Nazare, T. Bui and J. Collomosse, "Everything you wanted to know about Deep Learning for Computer Vision but were afraid to ask", 30th SIBGRAPI Conference on Graphics, Patterns and Images Tutorials, 2017.

[3] Swain S., Banerjee A., Bandyopadhyay M., Satapathy S.C. (2021) Dimensionality Reduction and Classification in Hyperspectral Images Using Deep Learning. In: Bandyopadhyay M., Rout M., Chandra Satapathy S. (eds) Machine Learning Approaches for Urban Computing. Studies in Computational Intelligence, vol 968. Springer, Singapore. https://doi.org/10.1007/978-981-16-0935-0_6.

[4] M. E. Paoletti, J. M. Haut, J. Plaza and A. Plaza, "Deep learning classifiers for hyperspectral imaging: A review", ISPRS Journal of Photogrammetry and Remote Sensing 158, 2019.

[5] P. Bhargavi and S. Jyothi, "Big Data and Internet of Things for Analysing and Designing Systems Based on Hyperspectral Images", Environmental Information Systems: Concepts, Methodologies, Tools, and Applications (pp. 621-641), IGI Global, 2019.

[6] H. Sajjad and P. Kumar, "Future Challenges and Perspective of Remote Sensing Technology", Applications and Challenges of Geospatial Technology, Springer, Cham, 2019.

[7] P. K. Garg, "Internet of Things and Remote Sensing", From Visual Surveillance to Internet of Things, Chapman and Hall/CRC, 2019.

[8] J. H. Schurmer, Air Force Research Laboratories Technology Horizons, Dec 2003.

[9] A. Fejjari, K. S. Ettabaa and O. Korbaa, "Chapter 12 Feature Extraction Techniques for Hyperspectral Images Classification", Springer Science and Business Media LLC, 2021.

[10] F. Mei, C. Zhao, L. Wang and H. Huo, "Anomaly Detection in Hyperspectral Imagery Based on Kernel ICA Feature Extraction", Second International Symposium on Intelligent Information Technology Application, pp. 869–873, 2008.

[11] L. Mou, P. Ghamisi and X. Zhu, "Deep Recurrent Neural Networks for Hyperspectral Image Classification", IEEE Transactions on Geoscience and Remote Sensing, PP. 1–17, 2017.

[12] W. Hu et al., "Spatial–Spectral Feature Extraction via Deep ConvLSTM Neural Networks for Hyperspectral Image Classification", IEEE Transactions on Geoscience and Remote Sensing, vol. 58, no. 6, 2020.
[13] Banerjee A., Swain S., Bandyopadhyay M., Rout M. (2021) Extraction of Information from Hyperspectral Imaging Using Deep Learning. In: Bandyopadhyay M., Rout M., Chandra Satapathy S. (eds) Machine Learning Approaches for Urban Computing. Studies in Computational Intelligence, vol 968. Springer, Singapore. https://doi.org/10.1007/978-981-16-0935-0_3.
[14] Z. Lin, Y. Chen, X. Zhao and G. Wang, "Spectral-spatial classification of hyperspectral image using autoencoders", 9th International Conference on Information, Communications & Signal Processing, pp. 1–5, 2013.
[15] L. Zhu, Y. Chen, P. Ghamisi and J. A. Benediktsson, "Generative Adversarial Networks for Hyperspectral Image Classification", IEEE Transactions on Geoscience and Remote Sensing, vol. 56, no. 9, 2018.
[16] Hyperspectral Dataset Source: http://www.ehu.eus/ccwintco/index.php? title=Hyperspectral_Remote_Sensing_Scenes

7

Artificial Intelligence and IoT for Smart Cities

Mukta Sharma

Associate Professor, HOD, Department of CS & IT, TIPS, India
Email: m.mukta19@gmail.com

Abstract

The benefits and enchantment of living in the 21st century are numerous. There are multiple technological pillars involved in building sustainable smart cities. This chapter will shed light on artificial intelligence (AI) and the Internet of Things (IoT) primarily. This offers new opportunities for collaboration in areas such as transportation, healthcare, industry, and the creation of digital infrastructure. Visualizing that now, no one gets stuck in the traffic congestion, while going to the office or for shopping. Every day, your doctor gets complete details about your daily sleep, heart rate, pulse rate, daily footsteps, calories burned, intake of calories, etc. Envision that seating in a different part of the world or just after leaving from your office and before reaching home, you want to make the room temperature apt for which you want to switch on your air conditioner/thermostat. Similarly, controlling all the appliances of the house is at your fingertips. AI and the Internet of Things (IoT) are two technologies that have the potential to transform cities into sustainable smart cities. This proposes new opportunities for collaboration in the areas of transportation, healthcare, industry, and the development of digital infrastructure.

The chapter will commence with a brief introduction about artificial intelligence, AI history, benefits of AI, limitations or challenges of AI, future scope, and solutions proposed by AI. The chapter will also highlight IoT, how IoT has actually made life easier and the devices to stay connected, and other advantages and challenges of AI. In this chapter, the author will discuss smart cities and various sectors like traffic management, agriculture, healthcare, etc., which have helped the cities get converted into smart cities.

7.1 Introduction

AI and IoT are two of the most disruptive technologies today, hastening the pace of business innovation. In this chapter, we will look at how the combination of these technologies can help not only with the development of new business models but also with the development of smart cities.

7.2 Artificial Intelligence

"It is the machine's ability to imitate human behavior." [28, 29]

The most basic definition of Artificial Intelligence would be "the intelligence demonstrated by machines" (Wikipedia, 2020). Machine intelligence is referred to as AI. Intelligent machines can improve performance and capabilities by predicting and analyzing [27]. It mimics human cognitive functions, such as problem-solving and learning. The application of artificial intelligence is expanding, and the number of jobs done by machines has almost doubled in recent years. Consider artificial intelligence to be an attempt to replicate or simulate human intelligence in a machine for better understanding. This enables them to complete tasks and make informed decisions based on the instructions and experience they have received. Artificial intelligence is a broad field of computing. It entails the process of creating smart machines and putting them to use in a variety of practical applications. Leading textbooks in this field define it more precisely as automatic thinking, humanization, and/or rational acting [34]. In general, the field of artificial intelligence seeks to advance science and intelligent engineering, with the aim of creating machines with human characteristics. This includes creating machines with a wide range of inspiring human abilities, such as communication, perception, planning, reasoning, the representation of knowledge, the ability to move and manipulate objects, and the ability to learn. AI uses tools and techniques from a variety of disciplines to solve problems, including probability, mathematics and statistics, philosophy, psychology, symbolic computing, linguistics, search and optimization, game theory, etc. [14, 17].

7.3 Artificial Intelligence History

Artificial intelligence (AI) is an old stream, approximately 50+ years old, collection of science, technology, and theories (including statistics, mathematical logic, computer science, probability, and computational neurobiology) designed to imitate human cognitive abilities. Its development began in the middle of World War II and was inseparable from computing, enhancing computers performance, and allowing it to process complex tasks that could only be entrusted to humans in the past [25, 30, 31].

Artificial intelligence originated in 1943 when Warren McCulloch and Walter Pits completed the first work by proposing a model of an artificial neuron. At the end of 1949, Donald Hebb showed an updated rule to modify the strength of the connections between neurons by proposing Hebbian learning, as a name to his rule.

Despite the fact that John von Neumann and Alan Turing did not develop the phrase "artificial intelligence" until 1950, they were the forefathers of technology. In the 19th century, they transitioned the computers to process and store the data, shifted from decimal logic (handling numbers from 0 to 9) to binary (based on Boolean algebra, handling 1 or 0). As a result, the two researchers formally identified the existing computer architecture and demonstrated that it is a general-purpose machine capable of executing programming material. The British mathematician Alan Turing first broached the issue of machine intelligence in his renowned article, "Computing Machinery and Intelligence," published in the 1950s in which he described a form of "imitation game." He imagined a computer that could communicate and respond by typing the messages so properly that it should be difficult to infer whether the person is talking to another person or a machine is conversing with the human. The Turing test can be used to assess a machine's capacity to demonstrate intelligent behavior comparable to that of a person.

In a summer symposium at Dartmouth College in 1956, John McCarthy of the Massachusetts Institute of Technology (MIT) coined the term artificial intelligence, which was later defined by Marvin Minsky (Carnegie Mellon University). According to Minsky "The creation of computer programmes concentrates on activities that humans currently accomplish more successfully because they demand high-level mental functions, such as critical thinking, perceptual learning, and memory organization." Since there was high enthusiasm for AI at the time, many high-level computer languages such as FORTRAN, LISP, and COBOL were invented.

Stanford Cart, a remote-controlled TV mobile robot, was invented in 1961, and in 1979, it successfully traversed a room full of chairs without human intervention and could be considered as the earliest example of the self-driving car. Between 1964 and 1966, Joseph Weizenbaum created the first chatbot called ELIZA. In 1966, the first mobile robot Shaky was born. Researchers emphasize the development of algorithms that can solve mathematical problems. In 1972, Japan manufactured the first intelligent humanoid robot WABOT1. However, before the mid-1980s, progress slowed and the field experienced the so-called "artificial intelligence winter." After nearly

8 years, sometime in 1980, Stanford University hosted the first National Artificial Intelligence Conference. The expert system was created to imitate the decision-making ability of human experts. The first vision-guided Mercedes-Benz robotic van, equipped with sensors and cameras, achieved a speed of 39 miles per hour (63 km) on roads with no traffic. This car was designed by Ernst Dickmanns.

Built under the leadership of the National Conference held at Stanford University in the 1980s, nearly 17 years later, IBM Deep Blue was launched in the year 1997 and defeated the world chess champion, Garry Kasparov, and became the first computer to defeat humans. Furby is a new pet household robot, invented in 1998 by Dave Hampton and Caleb Chung. Artificial intelligence (AI) entered the home in the form of the Vacuum Cleaner Roomba in 2002. AI-first appeared in the business world in 2006. Later, IBM's Watson demonstrated its ability to understand natural language in 2011, and it was very fast. Watson wins Jeopardy to solve puzzles by quickly solving all the complicated questions and riddles in the quiz show. Apple launched Siri in 2011, which is an integrated voice-controlled personal assistant for Apple users. In 2012, Google launched an Android application function called "Google Now" that can provide users with information in the form of predictions. It has the potential to recognize, learn to distinguish, and predict the correct object. Alexa was launched in 2014 by Amazon as a home assistant. "Eugene Goostman," a chatbot, won the infamous "Turing Test" contest in 2014. In 2016, Hanson Robotics developed a humanoid robot, Sophia, which is considered as the first "citizen robot." Google also launched Google Home in 2016, which uses artificial intelligence as a "personal assistant" to help user's complete tasks, such as memorizing tasks, creating appointments, and using their voice to learn information. IBM "Project Debater" debated complex topics with two debate experts, and its performance in 2018 was admirable. In the year 2018, Google announced Duplex, a virtual assistant for conversing naturally, over the phone in order to complete "real-world" tasks. For instance, Duplex had taken a hairdresser appointment on call, and the lady on the other end did not realize that she spoke to a machine.

AI has now progressed to a remarkable level, using technologies like deep learning, big data, and data science which are topics of great interest, and many kinds of research are going on for the same. The future of artificial intelligence is very exciting, and in the coming years, there will be highly intelligent machines. Companies such as Google, Facebook, IBM, and Amazon are now collaborating with AI to create amazing devices.

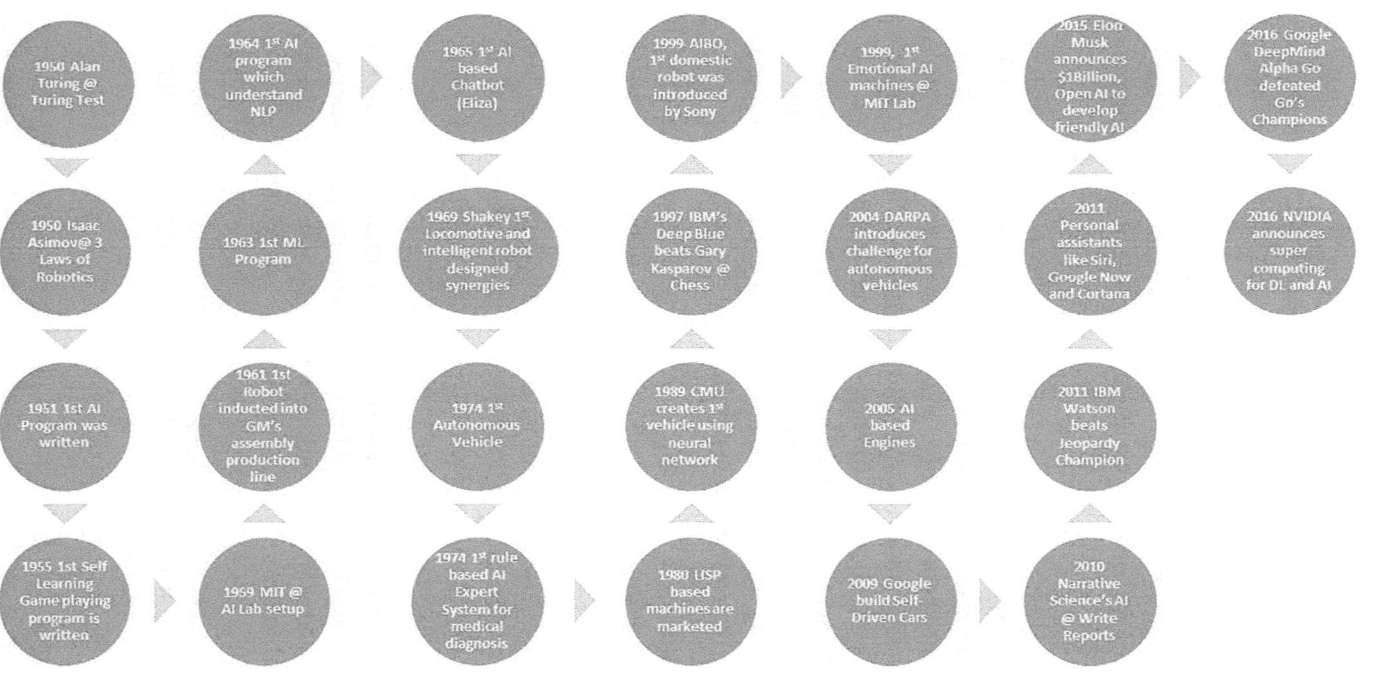

Figure 7.1 History of AI [36].

7.4 Benefits of AI

- **Human error is reduced/eliminated:** Artificial intelligence is making machines imitate human beings for which machines are trained and tested for accuracy and precision. Accuracy and precision gives artificial intelligence the capability of reducing errors significantly. The judgments made by AI in each stage are based on the information obtained earlier and a set of algorithms. If properly programmed, these errors can be avoided.

  This could prominently be seen in healthcare, and it deals with human lives and the healthcare companies are relying heavily on AI. One such example is detecting cancer in the nascent stage which can make the difference between the life and death of the cancer patients. AI has been proven to be more accurate and effective than humans in detecting new deep learning results for breast cancer and lung cancer screening; it is more accurate than digital mammography at detecting the disease. According to Tuitt, "recent research has demonstrated that AI is capable of picking up on small alterations that could diagnose certain tumors sooner than a traditional radiologist."

- **Automation**: One of the most commonly touted benefits of AI technology is automation, which has a big impact in almost all domains, may it be transportation, service industries, communications, consumer goods, and many others. Automation enables for more efficient use of raw materials, greater product excellence, shorter lead instances, and superior protection in diverse industries, similarly to growing manufacturing prices and productivity. Automation can also assist in the liberation of resources that can be put to better use.

- **Robots will take the risk instead of humans and can work 24 × 7:** Another key benefit of AI is that it allows people to delegate numerous risks to AI robots. According to Peter Scott, a NASA computer contractor and futurist, Artificial intelligence (AI) can make workers safer by allowing devices or robots to handle dangerous activities like dealing with nuclear waste (hazardous materials) or bomb disposal. It is also beneficial in minimizing the risks on health grounds that come with employment that involve repetitive movement, exposure to chemicals, or driving in dangerous situations.

  Machines with metal bodies or robots can cope with the risky situation for us as they are resistant in nature and can withstand adverse atmospheres. Situations like going to space or exploring the deepest parts of the oceans mining for coal and oil, to defusing a bomb, and many such difficult tasks. Furthermore, they are less prone to wear and tear and can deliver accurate work with higher responsibility.

Many research works have shown that human beings' productivity stays for roughly 4–5 hours per day and they frequently need to take breaks. AI, on the other hand, can work eternally without taking a break. They think far faster than humans and are capable of performing numerous tasks with great accuracy at the same time. According to Scott, "AI will eliminate the need for humans to perform monotonous, uninteresting, or unhealthy tasks such as driving to telemarketing."

AI-designed voice bots or chatbots have enhanced the customer's experience. Almost all large organizations utilize digital assistants to deal with their customers, considerably lowering the demand for human resources. The chatbots are outperforming humans and it is difficult to infer whether you are interacting with a machine or human being [11, 24].

- **New inventions leading to better decision making**: AI is driving a slew of new inventions in nearly every field that will help humans solve the vast majority of complex challenges. Combining AI with other technologies will make machines smarter and allow them to make accurate and smart decisions much faster than humans. Humans consider a variety of aspects when making decisions, both emotionally and practically, whereas an AI-powered computer works on what it is taught to do and produces results faster and without biases.

7.5 Limitations or Challenges of AI

AI has numerous advantages, but every coin has two sides. Similarly, AI also faces a few challenges, which are explained in the following.

- **Expensive:** Developing a machine that can imitate human intelligence is no minor effort. The high level of engineering that goes into creating AI-based machines, computers, and other devices comes at a high price. In addition, the excessive cost of repair and maintenance can go into the thousands of dollars. AI also necessitates the most up-to-date hardware and software to stay current and fulfill the most recent criteria, making it highly costly.

Do you know that Apple paid almost $200 million to acquire its virtual assistant Siri? Furthermore, the fact that Amazon paid $26 million in 2013 for Alexa demonstrates the high cost of AI implementation [6].

- **AI lacks creativity**: One of the biggest flaws of AI is that it cannot be trained to think outside the box. AIs lack creativity and imagination. Human brains are recognized for their high emotional quotient and great sensitivity. In short, AIs can learn to be skilled robots, but they will never match the human brain's capabilities. Because, while skills can be learnt and mastered, abilities are innate and can only be polished; this is the case.

  AI learns and analyzes based on archived data. Based on the pre-fed data and archived data or experiences, AI machines learn over time, but they cannot be creative in their approach. Let us see with an example, the bot Quill, which writes Forbes earnings reports, as a classic example of AI. Only data and information that the bot has already provided are included in these reports. Human touch is found lacking in the Forbes article when written by bots though the article looked impressive [6, 8].

  AI does not grow with experience: It has been observed and debated that human cognitive power improves with age and experience. AIs, on the other hand, are machines that do not improve with experience, but rather wear and tear over time. Machines are incapable of adapting to changing conditions. Machines cannot adapt to changing environments. Repetitive work with constant input is the underlying premise on which AIs are formed. As a result, the AIs must be re-evaluated, re-trained, and rebuilt every time the input changes.

- **Risk of unemployment:** According to a McKinsey Global Institute estimate, intelligent agents and robots would replace 30% of the world's existing labor by 2030.

  As a result, it is impossible to rule out the possibility that AIs may lead to less human intervention, thereby disrupting job standards. Most businesses are now using some amount of automation in order to replace low-skilled workers with machines that can perform the same tasks more effectively.

- **No emotions and ethics:** Machines are unquestionably superior to humans in terms of productivity, but they cannot replace the emotional connection that bonds the team together. Machines cannot be emotionally tied to humans in the same way that humans can be emotionally bound to other humans, which is an important characteristic in team management. By now, the readers are familiar with a chatbot, where, initially, the machine is answering the queries and, for any human intervention, a ticket is generated and the call is diverted to the customer care executive to handle and attend to the customer. As customers can answer out of prefixed question sets, they can also listen and show empathy if the client is in pain or stress, and one cannot expect emotional tinge from the machines.

7.6 Applications of Artificial Intelligence [14, 21]

Machine intelligence that replicates human intelligence is known as artificial intelligence. According to Tractica, the global market is expected to generate $118.6 billion in revenue by 2025. AI usage has increased by 270% in the last few years, as per the research done at Gartner, indicating a clear indication of future growth. According to one study, 87% of companies that have implemented AI are using it to improve email marketing. In over 75 countries, artificial intelligence is being used for surveillance. Now, let us explore how AI is being used in different domains.

- **AI in business**: It attracts and personalizes customers, keeping them engaged by showing their preferences and interests based on their profile data. This will increase the company's brand loyalty and revenue generation. It also supports visual search; so if users are searching for something, they can upload an image of the article and search the data accordingly.

  For instance, virtual assistants and chatbots are AI-powered helpers that improve the online purchasing experience, and natural language processing (NLP) approaches aid in responding to consumers. Chatbots are already being used by many apps and websites to interact with users and provide personalized assistance in answering their questions and keeping them engaged. They are also employing AI to gather feedback or reviews on their most recent purchase. Adaptive intelligence, chatbots, and automation all contribute to smoother business processes.

  AI also not only the customers but also the business houses by aiding in forecasting product demand and improving decision making, resulting in higher revenue and more loyal customers. Questions such as (what item to stock, when, etc.)

Many buyers choose to acquire a product or service based on the feedback and reviews given by other customers. AI can assist in detecting and dealing with fake reviews. AI handles fake reviews by identifying and handling them in the process of improving customer's trust in your brand and products. Companies like Amazon, eBay, Alibaba, etc., are already using AI.

- **AI in robotics:** AI-enabled robots will be able to navigate their paths with real-time updates. It uses this path to detect obstacles and plan its route. Sophia, the robot, is a real-life example of this application. Robot Sophia dubbed a "humanoid," which is a hybrid of a human and a robot with abilities from both.
- **AI in healthcare:** AI is being used by many organizations and medical care centers to save human lives. The role of AI in healthcare enables machines to interpret a patient's medical history and predict potential diseases to which the individual may become vulnerable in the coming years. Furthermore, AI has aided in drug discovery and the development of medicines that can treat harmful or even fatal diseases. The best example could be the vaccination for COVID-19 – not only this but also predicting the second wave and many more [26].

AI-powered audit systems can aid in the prevention of prescription errors and the correct identification of diseases. Prescriptive analysis of patient data can be performed to ensure that the case is prioritized and triaged in real time.

Medical imaging and diagnostics: To ensure early detection of chronic illnesses, artificial intelligence (AI) can assist in the processing of lab and other medical data. In advanced medical imaging, AI can also be used to analyze and transform images.

For a better understanding, let us look at a few real-life examples. Coala Life, a company that has developed a digitalized gadget for detecting heart problems, is one example. Another example, Cambio Health Care created a clinical decision support system for stroke prevention that can notify doctors when a patient is at risk of a heart attack. Aifloo, meanwhile, is developing a method to monitor how individuals are doing in home care, nursing homes, and other settings. The nicest benefit about implementing AI in healthcare is that you will not have to develop any new drugs. Simply by correctly administering an existing drug, you can save lives [39]. AI has had a significant impact on the industry. IBM Watson Health is one medical firm that has actively incorporated AI into the healthcare industry. It aids in research, data analytics, and the delivery of medical solutions to clients.

- **AI in banking and finance:** AI in banking is expanding faster and many banks already have AI-based systems to provide customer care, detect anomalies, and detect credit card fraud, and this trend is expected to continue. Since its debut, HDFC Bank has handled over 3 million client questions, interacted with over 500,000 unique customers, and held over a million discussions through Electronic Virtual Assistant (EVA). Eva can take data from thousands of sources in under 0.4 seconds and deliver straightforward replies.

  One of the most important difficulties that e-commerce enterprises confront is credit card fraud. By taking into account usage trends, AI can help lower the possibility of credit card fraud.

  Likewise, Nomura Securities, Japan's largest brokerage business, has been unwillingly pursuing one goal: utilizing computers to assess the insights of experienced stock traders. After years of development, Nomura is about to debut a new stock trading system. The computer in the new system keeps a huge amount of pricing and trading data. It will make decisions based on the information it has access to. It may, for example, conclude that present market conditions are comparable to those of two weeks ago and forecast how share values will change in a few minutes. This will allow traders to make better trading selections based on the projected market price.

- **AI in agriculture:** The provision of precision statistics has extended the production and implementation of AI in agriculture. AI in agriculture has almost eliminated the obstacles to traditional farming, allowing farmers to be future-ready. AI creates a statistics repository for farmers to rely upon through collecting and analyzing records. The agriculture industry enthusiastically and freely includes AI into its practices to enhance typical performance. Artificial intelligence is used in a variety of methods on farm equipment. In numerous areas of an area, ranch apparatus will plant diverse densities of seeds and practice numerous quantities of manure. It allows in awaiting warnings for planting, bug manipulations, shower insecticides, inundating, and object valuing which will assist Indian ranchers to increase their efficiency and pay [37].

AI is used extensively in agriculture. For example, See & Spray, a robot designed by Blue River Technology, monitors and precisely sprays weedicide on cotton plants using computer vision technology such as object detection. Herbicide resistance can be avoided by precision spraying. Artificial intelligence can be used to detect soil flaws and nutrient deficits. Plantix is an application developed by PEAT, a Berlin-based agricultural tech start-up, which uses photographs to identify probable flaws and nutrient deficits in the soil.

The app for image recognition scans photographs taken using the user's smartphone camera for potential defects. Users are then given soil restoration strategies, ideas, and other potential solutions. The software can detect soil faults and nutritional deficits. Weeds can be detected using artificial intelligence. Agriculture bots can harvest crops in greater volume and at a faster rate than human labor.

- **AI in education:** Many times, teachers are unaware of the gaps in their lectures and educational materials that students may encounter. This can cause students to be perplexed about certain concepts. With AI, the system alerts the teacher and informs the problem. It sends students a personalized message with hints to the correct answer. As a result, this helps to fill in any explanation gaps that may occur in courses. It also ensures that all students are constructing the same conceptual foundation.
- **AI in gaming**: AI has the potential to be used in gaming. AI machines can play strategic games like chess, in which the machine must consider a large number of possible locations. AI can be used to save labor costs and to generate levels, maps, textures, weapons, characters, etc. Artificial intelligence (AI) can be utilized to create intelligent, human-like NPCs that interact with gamers. To improve game testing and design, it can forecast player behavior.

  The AI's actions in chess are designed to be unpredictable and make the game extremely difficult. This means that players must swap strategy frequently and never stay in the same position for long enough for the AI to learn from past generations' failures.
- **AI in autonomous vehicles:** Self-driving cars are built using artificial intelligence. AI can be utilized in conjunction with the cloud services, GPS, camera, control signals, and radar, to operate the vehicle. By adding functions like blind-spot monitoring, emergency braking, and driver-assist steering, AI can improve the in-vehicle experience.

The following are some features that are used by self-driving cars:

a. The direction of travel is modifies based on known traffic conditions to find the quickest route.

b. When the automobile runs out of fuel, it navigates to a gas station or charging station automatically.

c. Voice recognition is included to improve communication with passengers.

d. Such technologies include natural language interfaces and virtual assistive technologies.

Tesla's self-driving car is an example of an autonomous vehicle. It makes cars that can recognize objects and navigate without the need for human involvement using computer vision, image detection, and deep learning.

- **AI in social media:** Millions of individuals use social media platforms like Instagram, Facebook, Twitter, and Snapchat to keep connected online. Artificial intelligence influences the vast majority of decisions; everything, from notifications to upgrades, is curated by AI. It takes into account all previous web searches, behaviors, interactions, and much more. As a result, when users visit these websites, the data is stored and analyzed, and the user is provided with a personalized experience. It handles billions of user profiles in an efficient manner. AI can analyze large amounts of data to identify the most recent trends, hashtags, and user requirements. AI algorithms are used for sentiment analysis, also known as opinion mining, and are a powerful tool for building smarter products. Sentiment analysis and machine learning are used by social media monitoring apps and companies to gain insights about mentions, brands, and products.

- **AI in entertainment:** AI-powered applications such as Netflix and Amazon are widely used. These services use ML/AI algorithms to make recommendations for programs or shows.

MuseNet is an AI-powered music composition system that will use ten distinct instruments and styles ranging from country to Mozart to the Beatles. MuseNet was not specifically intended to understand music; instead, it self-taught itself to recognize rhythm, style, and patterns of harmony.

Another remarkable artificial intelligence product is WordSmith, a content automation platform. Wordsmith is a natural language creation technology that may help you turn your facts into compelling stories. WordSmith is used by tech giants like Microsoft, Yahoo, and Tableau to create billion pieces of content per annum.

- **AI in space exploration:** Artificial intelligence is used not only on earth but also in space. Whether it is Mars missions or satellite installations in the exosphere, artificial intelligence is steadily progressing toward the exploration of outer space.

  The examination of enormous amounts of data is always required in space research and discoveries. Artificial intelligence and machine learning are the most effective ways to handle and process data of this magnitude. After significant research, astronomers utilized artificial intelligence to sift through years of data collected by the Kepler telescope to locate a distant eight-planet solar system.

  Artificial intelligence is also being used for NASA's mission to Mars, the Mars 2020 Rover. The AI-powered AEGIS Mars rover is already on the red planet. In order to perform study on Mars, the rover is in charge of autonomous camera targeting. Artificial intelligence is being employed in space exploration in a variety of ways, including map-building, satellite navigation, and location tracking.

7.7 IoT

The internet of Things (IoT) is an idea that has only recently entered our everyday lives. It is all around us: connected cars on the road, home automation devices in our homes, smart office sensors embedded in our offices, and fitness trackers worn on our bodies.

IoT encompasses all that connects to the internet, including your smartphone, laptop, tablet, and desktop, is included in the Internet of Things. However, the phrase is widely used in a slightly narrower meaning, referring to various devices that may communicate with one another – smart speakers, lights, heating systems, plugs, refrigerators, cars, and so on – presuming that smartphones and PCs are already internet-connected. The Internet of Things, on the other hand, technically refers to any device that connects to the internet and has an on/off switch, including mobile phones. The Internet of Things (IoT) is all about connecting things [13, 23].

Let us look at an example of how IoT works: Let us pretend someone is driving back home from work. The car is linked to a smartphone, which, in turn, is linked to a smart refrigerator at the user's residence. The refrigerator sends a text message to the phone alerting the out of milk status, which is shown on the dashboard of the car, along with the location of the nearest grocery store and a map with directions. The store shelves are also linked, and a notification appears on the dashboard of the car indicating that the selected brand is in stock [14].

The Internet of Things (IoT) is viewed as a technology innovation and monetary wave in the worldwide information data industry after the internet. The IoT is a clever system, which associates everything to the internet for the sole purpose of exchanging data information, which has been communicated through agreed sensor devices protocol. IoT accomplishes the objective of intelligent tracking, monitoring, locating, identifying, and managing things [38].

7.8 History of IoT

The IoT is a network of physical objects – "things" – embedded with sensors, software, and other technologies that enable them to communicate with other devices and systems over the internet.

The first invention of the electromagnetic telegraph by Baron Shillings, Russia gave the concept of connected devices back in the year 1832. The telegraph allowed two machines to communicate directly by sending electrical signals back and forth. Samuel Morse sent the first public telegraph message in 1844. However, the true Internet of Things history began in the late 1960s with the invention of the internet, a critical component that grew rapidly over the next decades.

Although it may seem difficult to believe, the first connected device was a Coca-Cola vending machine operated by local programmers at Carnegie Melon University. They used micro-switches in the machine and an early version of the internet to determine whether the cooling device was keeping the drinks cold enough and whether Coke cans were available. This discovery sparked additional research in the field as well as the development of globally interconnected machines.

In the year 1990, John Romkey introduced the first Internet of Things device. He created the first smart toaster, which can be controlled remotely over the internet. He demonstrated his invention at the INTEROP conference.

In 1999, while working for Procter & Gamble, Kevin Ashton invented the term "Internet of Things." Ashton was interested in learning more about RFID, or radio-frequency identification, because he was working on supply chain optimization. RFID, as the name implies, uses radio frequency to identify, locate, track, and interact with objects and people. It is a radio frequency tracking system that can store smart labels ranging from serial numbers to brief descriptions to entire pages of data.

The invention of IoT, which uses IP address (IPv6) that goes beyond the constraints of IPv4, has transformed the internet world by connecting a massive number of smart linked gadgets or devices, estimated to be over 70 billion or more. The Second Economy, also known as the Industrial Internet Revolution, is thriving as a result of this innovation. It will create a massive market with a variety of services, with a market size estimated in the trillions of dollars. This industry has the potential to be a success, but only if security and privacy concerns are addressed before this massive procedure is implemented globally [16].

Dublin was named the first Internet of Things (IoT) city in 2014. Smart Dublin was a city-improvement initiative that made use of Internet of Things (IoT) devices. A new carbon-neutral stadium, hundreds of smart bins, flood-monitoring sensors, and city sound monitoring sensors are among the enhancements. Dublin hoped that by launching this initiative, they would inspire tech innovation and entrepreneurship in their city, paving the way for future growth.

7.9 Advantages of the Internet of Things (IoT)

- **Easy access and better monitoring:** Monitoring or accessing the information easily is the primary and most significant benefit of IoT. One can now easily access critical information in real time from (almost) anywhere. A smart device and an internet connection is all that is required to determine the exact quantity of supplies like to know when your printer ink is running low or when your refrigerator has low milk or vegetables. This will save time and users can plan accordingly also if someone wishes to know the air quality in your home. Keeping track of when products expire will also improve safety.

  Google Maps is an apt example of using IoT, as it helps us find the location and also navigate the other routes to reach the same destination, along with precise details about traffic congestion and a tentative time to reach, rather than asking someone on the street. Factual information is easily accessible at your fingertips, even from the most recent scientific research or business analysis.

- **Enhanced productivity of staff and reduced human effort:** With all of this data coming in, one can complete a slew of tasks at breakneck speed. Smart workplaces automate monotonous operations, allowing employees to concentrate their time and effort on more complicated jobs that demand personal abilities, particularly out-of-the-box thinking, thanks to the Internet of Things.

When IoT is used in inventory control, it saves time and labor in the entire process because IoT provides accurate data and precision. When less labor is used, revenues will eventually rise.

One can quickly set up and roll out new products and services, thanks to IoT capabilities. IoT devices will gather the data and help in predicting the buying pattern of the customers which will accelerate the planning and will deliver the product which could generate better revenue. Combine IoT data with historical data to identify areas for new work in advance [4, 32].

- **Comfort:** Driverless cars, also known as self-driving cars, are a fantastic option, especially after a long tiring day at work. Imagine waking up in the morning and finding that the tea/coffee is ready to go according to the person's preferences; all thanks to the Internet of Things.
- **Adapting to new standards:** Despite the fact that the Internet of Things is constantly evolving, its changes are minimal in comparison to the rest of the high-tech world. Without IoT, it would be a struggle to keep up with all of the latest updates.
- **Improved time management:** One can keep themselves updated with the latest news available on phones while traveling to home or can check a blog about any specific hobby or shop from an online store; one can do almost everything from the palm of our hands. Overall, it is an ingenious time-saving tool.

  IoT primarily assists people in their daily lives by allowing their devices to communicate with one another in an efficient way, by saving and protecting energy and cost and eventually providing monetary benefit to the end-user.
- **Automation:** Another significant benefit of smart device interconnection is automated control over a variety of operational areas, such as tracking the shipment, inventory management, spare parts management, and many more. RFID tags and a network of sensors are good examples used to track the location of equipment, goods, and also animals in wildlife sanctuaries [10].
- **Safety:** When computers keep track of the food, medicine, and expiry dates, it becomes safer. Working in a safe environment makes the

company more appealing to stakeholders, may it be investors, clients, employees, or company trading partners; this will enhance brand reputation and trust of the company. Smart devices also lower the risk of human error at various stages of the business process, resulting in increased safety. An IoT network of motion sensors, surveillance cameras, and other monitoring devices can also be utilized to safeguard an enterprise's security and stop theft and corporate espionage.

- **Enriched customer service and retention:** Businesses can better understand customer expectations and behavior by using user-specific data received from smart devices. IoT also increases customer service by providing post-sales follow-ups such as automatic tracking and reminders of required equipment maintenance after a predetermined duration of usage, the expiration of warranty periods, and so on.
- **Health benefits**: IoT helps in the monitoring of patients. With the help of IoT, patients can easily be diagnosed and given proper treatment because patient bio-data is being stored for future observation and analysis.

7.9 Disadvantages of Internet of Things (IoT)

- **Security concerns:** The most common impediment to the overall development of IoT is insufficient security measures. With so much data being transmitted, if there is ever a security breach of data, it means people become vulnerable to attack due to the Internet of Things. Data leaks are always a source of concern since smart gadgets collect and send sensitive data that, if exposed, might have severe effects. Identity theft, equipment, or products sabotaging loss of corporate secrets, and other fatal repercussions are all possible.

  Imagine a scenario where someone can easily hack into your home TV and the hacker or intruder can listen to your conversations or can keep a watch on all the activities which can be a big threat to us [35]. Imagine a hacker, hacking into a self-driven car, where people can easily be assassinated or murdered. So it is imperative to make sure there are safety measures in place for that not to happen [4, 10].
- **Implementation in business could be expensive initially:** The establishment of a broad network comprising multiple smart devices as well as the associated technical infrastructure, such as the power supply grid and the communication network, is required to implement IoT

infrastructure in a firm. Smart devices, first and foremost, require a reliable and sufficient supply of electricity, necessitating the construction of well-planned new infrastructure. It should include a sufficient number of surge protectors, UPS devices, and other ingress protection (IP) rated equipment.

The core aspects of the Internet of Things are the vast number of interconnections between diverse devices and access to the global network. As a result, IoT devices require an infrastructure for both wired and wireless continuous communication with high throughput, low latency, and constant internet connectivity. To reap the benefits, a business must first provide all of the required networking equipment, such as cables, hubs, routers, local data storage devices, and so on.

As a result, such a project will demand significant investments in order to build, maintain, and eventually grow the network developed in response to future needs. Despite the numerous advantages of IoT solutions, they take a long time to become profitable, and the financial rewards far surpass the initial deployment expenses.

- **Complexity:** While IoT appears to manage tasks with ease, it actually performs a number of complex operations behind the scenes. If the software makes an incorrect calculation by accident, it will have an impact on the rest of the process. Like one will have no idea how to deal with the incorrect temperature to control the thermostat at home. In the worst-case scenario, an error code in water dam software could result in a disastrous flood. An error in IoT is difficult to debug. The complexity of the Internet of Things means there is complexity for failures, which can effectively affect or slow down systems. The malfunction of an electric grid can cause a power outage. Someone can easily hack into your power grid just to bother you or discomfort you.
- **Need highly skilled people:** Responsible, experienced professionals who grasp the scope and potential ramifications of their activity are necessary for IoT solutions. In order to deploy, configure, maintain, and scale IoT solutions in a business, highly trained administrators are required, which might be difficult to locate and hire due to their respective high salaries. All staff who will be working with the smart device network should receive thorough training and comprehensive instructions. As a result, while the Internet of Things decreases the need for human resources, those who remain must be well-trained to prevent disturbing smart device operation and producing the "snowball effect."

- **Unemployment:** Non-educated, unskilled workers and helpers may lose their jobs as a result of the automation of daily activities. This has the potential to result in societal unemployment.

  With technology and the Internet of Things, taking virtually everything is a problem but can be overcome with education. Naturally, as daily activities become more automated, the demand for human labor, primarily workers, and less educated personnel will decrease.

7.10 Application of IoT

- **Smart homes:** Smart homes are one of the finest and most useful IoT applications, elevating convenience and home security to new heights. Though the Internet of Things can be employed at various levels in smart homes, the optimal solution combines intelligent utility systems and entertainment. Let us understand smart homes with a few examples which make the home a smart home such as the electricity meter with an IoT device that provides insights into your daily electricity consumption, your set-top box which allows you to record shows remotely, automatic illumination systems, advanced locking systems, connected surveillance systems, etc. [19].

  Being one of the best and most practical IoT applications, smart homes take home security and convenience to new heights. Though the Internet of Things can be employed at various levels in smart homes, the optimal solution combines intelligent utility systems and entertainment.

  IoT devices or gadgets are a piece of the bigger idea of home automation, otherwise called "demotics." Vast brilliant home systems use a main hub or controller to provide clients with a focal control for the majority of their devices or gadgets. These gadgets can include lighting, air conditioning, heating, security, and media system [20]. These devices are majorly controlled by a voice control mechanism. An example is Amazon's Alexa.

- **Wearables:** Wearable technology is a hallmark of IoT applications, and the healthcare industry is among the first few to use it. Virtual glasses, fitness bands like FitBits, smartwatches that track calories burned, monitor the heart rate, and GPS tracking belts are just a few examples of wearable technologies that can be found practically anywhere nowadays. These are compact, energy-efficient gadgets that collect and manage data and information about users using sensors, measurement and reading hardware, and software.

The Guardian glucose monitoring device is one of the lesser-known wearables. The gadget was created to help control diabetics. It measures glucose levels in the body by implanting a tiny electrode known as a glucose sensor beneath the skin and transmitting the data via radio frequency to a monitoring device.

- **Healthcare:** Doctors stay connected to patients through wearables or sensors and can observe patient's health from anywhere as the real-time data is collected through the wearable sensors. The Internet of Things aids in enhancing the patient care and preventing fatal happenings in high-risk patients by constantly observing key metrics and sending automatic alerts on their vital signs [12].

  Observing health and disaster from a distance or emergency notification systems can both benefit from IoT devices. From heart scanners and blood pressure monitors to complex devices capable of observing customized implants, these healthcare technologies or gadgets are available. Another example is the integration of IoT technology into hospital beds, resulting in smart beds equipped with unique sensors to monitor vital signs, blood pressure, oximeter, and body temperature, among other things, such as Fitbit electronic wristbands, pacemakers, or improved hearing aids [9, 12].

  Telemedicine, also known as telehealth, has not yet reached its full potential; despite this, it has a promising future. Video consultations with specialists, medical imaging, remote medical diagnosis and evaluations, and other IoT telemedicine applications are growing high.

- **Agriculture:** Agriculture is one of the businesses that have gained the most from the Internet of Things. The future looks bright with so many developments in agricultural gear. Drones are being developed for farm surveillance, drip irrigation, crop pattern interpretation, water distribution, and other purposes. These will allow farmers to produce a higher-yielding crop and address their concerns more effectively.

  The amalgamation of cloud services and remote sensors in agriculture helps gather crucial data relating to the ecological conditions – rainfall, humidity, temperature, wind speed, soil nutrients, and pest infestation, which are connected with farmland, can be utilized to enhance farming methods, make informed choices to enhance quality, and limit dangers and waste. The application-based crop or field checking additionally brings down the issues of overseeing crops at numerous areas or locations. For instance, agriculturists would now be able to identify which zones have been fertilized, also pesticides will be sprayed only where weeds are growing (or erroneously missed), and know more about the land condition as to whether it is dry or moist for anticipating the crop yield [37].

- **Transportation and traffic management:** The Internet of Things is contributing to the smart city by making a very valuable contribution by managing the traffic in cities. The IoT can aid the integration of control, communications, and information processing across different transportation system platforms [22].

  The facilities like smart traffic control, smart parking, vehicle control, road assistance, driverless cars, electronic toll collection (ETC) systems, and logistic and fleet management enhance the effectiveness of IoT in transportation systems.

  Users are using IoT on mobile phones, while using the applications like Waze or Google Maps, to monitor the traffic, as it also shows the different routes, and give proper information about time, congestion, distance, and estimate. Meanwhile, your phone sensors are collecting and sharing your data for these applications.

  - **Driverless cars:** There has been much debate about self-driving cars. Google and Tesla experimented with it, and Uber even developed a self-driving car prototype that it later abandoned. Technology needs to ensure better safety for passengers on the road as human lives are involved in it. The vehicles are connected to the cloud and data is continuously generated and collected via numerous sensors and embedded systems in the car; this can be used for decision-making. This technology will take a few more years to fully grow. The countries need to change their laws and policies, to see a promising future for driverless cars.

  - **Fleet management:** Sensors installed in fleet vehicles aid in the establishment of effective interconnectivity between vehicles, managers, and drivers. Accessing the software, one can gather, process, and organize the data. Receive instantaneous alerts to upkeep incidents that have not been noticed by the driver. The IoT helps in telemetry control and fuel savings, the reduction of polluting emissions to the environment, geolocation (route monitoring and identification of the most efficient routes), and performance analysis and can also provide valuable data to improve driving.

  - **Parking availability:** It has been observed that parking leads to traffic congestion especially in metropolitan cities. Palo Alto, San Francisco, is the first city of its kind to take a completely different approach to traffic. They realized that most cars on the streets circle the same block in search of a parking spot. That is the primary cause of traffic congestion in the city. As a result, sensors were installed at all parking spots throughout the city. The occupancy status of each spot is transmitted to the cloud by these sensors. That data can be consumed by a wide range of applications. It can direct drivers to the shortest path to an available parking space [39].

- **Smart supply-chain management:** Businesses have been using the concept of supply-chains for quite a long time now to track the goods while they are on the road, which is a common example. They are certain to remain in the market for the foreseeable future, as they are supported by IoT technology.

7.11 Smart Cities

[1] Cities have faced traditional challenges especially as the population is increasing in the urban areas; it is a matter of concern with regard to parking, traffic, crime, electricity, pollution, safety, and many more. Most of the cities have a plan to combat them by becoming smart cities. Cities can think, analyze, and make a decision or help the citizens to make a valid decision. Smart cities use technology very efficiently and seamlessly to enhance the quality of life; for instance, the commuters can reach the destination faster by avoiding congestion and by taking the shortest route.

A smart city is a platform for creating, deploying, and promoting sustainable development strategies in order to address urbanization's mounting problems. The majority of it is made up of information and communication technology (ICT). One of the most important aspects of constructing smart cities is the Internet of Things. The Internet of Things (IoT) is a network of physically connected things that exchange data to communicate with one another. Smart city technology is linked by the Internet of Things (IoT), and sensors, lights, and meters are a few examples of IoT devices that collect and analyze data.

Electrical signals are captured by electronic, thermal, infrared, and proximity sensors, which are then analyzed by humans or artificial intelligence. Sensors could be employed in a smart city to observe things like electricity usage, lighting, traffic, and weather. Artificial intelligence (AI) is a computer simulation that can make decisions in the same way as humans do. AI, for example, can count automobiles, people, or any other form of activity and track their speeds. Face recognition, license plate scanning, and the analysis of all satellite data highlight AI's ability to generate patterns for city planning.

A smart city, according to IBM, is one that utilizes all available networked data to better understand and regulate its operations while optimally using the resources.

As depicted in Figure 7.2, a smart city's four pillars are physical infrastructure, institutional infrastructure (including governance), social infrastructure, and economic infrastructure. In other words, a smart city aims to deliver the greatest services possible to all of its citizens, regardless of their social class, age, income level, gender, or other criteria [5, 18].

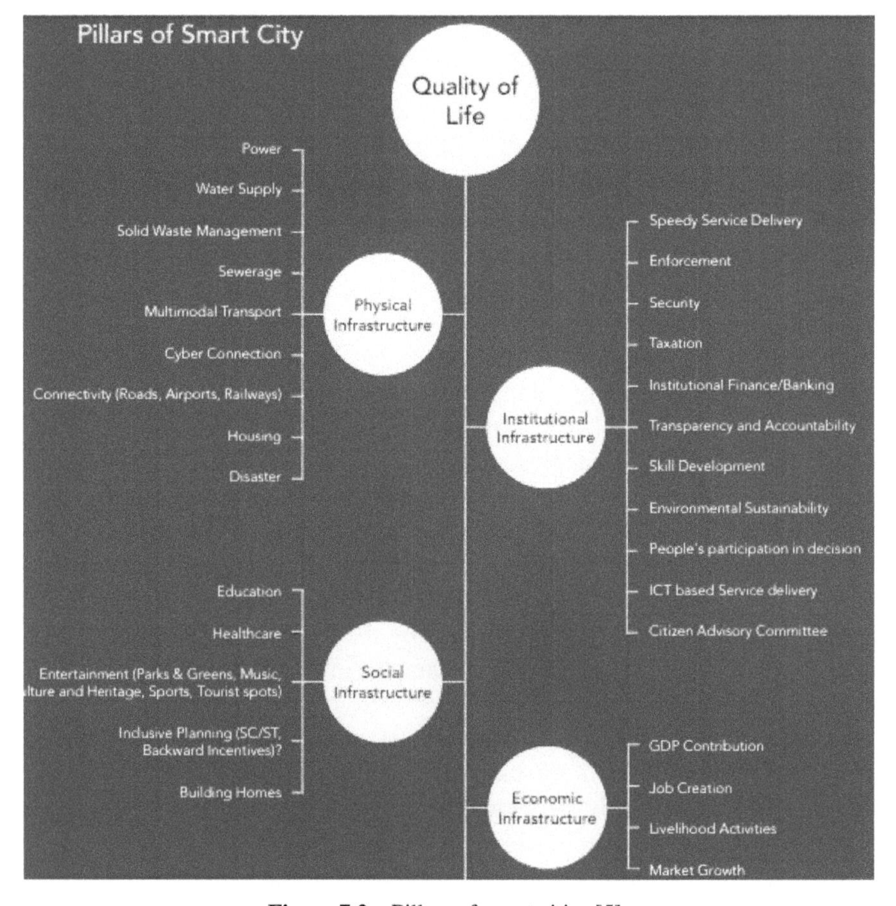

Figure 7.2 Pillars of smart cities [5].

- **Social infrastructure** refers to systems that work to develop human and social capital, such as healthcare, educational, and entertainment systems.
- **Physical infrastructure** refers to the city's physical infrastructure, such as its water supply system, sewerage system, transportation system, housing stock, energy system, sanitation facilities, solid waste management system, drainage system, and so on.

  The activities associated with a city's planning and management systems are referred to as **institutional infrastructure**. Cities require high-quality governance, as well as a strong local voice in decision-making. The guiding premise is usually "Governance by Incentives rather than Regulation."
- **Economic infrastructure**: A smart city must first identify its key capabilities, relative advantages, and impending economic activity in order to attract investments and construct the appropriate economic infrastructure for employment creation. Incubation centers, industrial parks, skill development centers, export processing zones, IT parks, trade centers, service centers, financial centers, services, logistics hubs, warehousing and freight terminals, mentorship, and counseling services are just a few examples.

7.12 How Do Smart Cities Work?

The smart city keeps on collecting data, processing it, analyzing the same data, and updating the users; for instance, cities data is collected via IoT sensors and video cameras and decision is taken based on the analysis. Let us learn more about the four steps taken by the cities to enhance the quality of life while facilitating economic progress:

a. Data acquisition: Data is collected in real time by smart sensors.
b. Data analysis: In order to have a better understanding of how the city's services and operations work, the data is analyzed.
c. Communication: Decision-makers are informed of the results of the data analysis
d. Action: Efforts to improve operations, asset management, and the quality of life of city people are made.

The information and communications technology framework collects actual data from joined assets, objects, and machines to aid decision-making.

Citizens, on the other hand, can use mobile devices, connected vehicles, and buildings to interact with smart city ecosystems. By connecting gadgets with data and city infrastructure, it is feasible to cut expenses, increase sustainability, and streamline elements such as energy distribution and trash collection, as well as minimize traffic jamming and increase air quality.

It senses the environment in order for the city operator to know when and how to react. A few actions can be completed automatically, such as contacting the nearest municipality or relevant authorities for pickup rather than waiting for a scheduled pickup. A smart city connects, protects, and improves the lives of its citizens by utilizing digital technology. Sensors, video cameras, social media, and other inputs act as a nervous system, which constantly provides feedback to the city operator and citizens to make informed decisions.

7.13 Advantages of Smart Cities [3]

- **More efficient decision-making with the help of real-time data:** Smart cities aid in the management of cities by allowing for more effective, data-driven decision-making. Efficient big data technologies and strategies could assist a city in identifying and staffing police in high-risk areas, forecasting and planning for citywide population growth, and identifying trends in citizen concerns, interests, and needs. This will open up a world of possibilities for better decision-making. As a result, inhabitants' lives are improved as prices are reduced and services are improved.
- **Better transport services:** Traffic congestion is a major problem for most of the city dwellers, and technology is proposing many solutions to it. For instance, public transportation like (buses, metro, trains, etc.) routes can be adjusted in real time based on demand, and intelligent traffic lights can help to reduce congestion and improve traffic flow. Residents can also be notified when it is safe to take public transportation after hours using smart technology. Riders of public transit can track their bus or train whereabouts and alter routes as needed in numerous places.
- **Safer cities are safer:** Smart cities are well connected with Wi-Fi, IoT, sensors, and surveillance cameras, which helps in providing a more secure environment. Smart cities enhance resident safety and reduce incident response times. By providing precise and timely information, license plate recognition, networked crime centers, and cameras can

all assist law enforcement. Many new technologies like face recognition, fire and smoke alarm capabilities, and even door locking and unlocking are helping reduce criminal risk. To make communities safer, these technologies work together. Utilizing technical breakthroughs and establishing private/public partnerships can help to reduce criminal activities.

- **Infrastructure improvements:** Normally, the city's infrastructure involves roads, bridges, and buildings, among other infrastructure components, necessitating massive investments in maintenance and repair as they age. The technology can identify problem areas before they cause a failure by predictive analytics. The sophisticated sensors detect structural changes such as tilts and cracks in buildings and bridges. The sensors will send out alerts when inspections or maintenance are required. This way, the cities help save money and lives by preventing preventable infrastructure failures [15, 33].
- **Citizen and government engagement:** Citizens, nowadays, expect cities to provide reliable and user-friendly digital services. In many sectors of life, collaboration tools, mobile applications, modern and user-friendly websites, self-service portals, and convenient online accounts have become the new norm, and people demand no less from their city. Digital services are spreading in communities, making smart cities a fantastic place to live while also encouraging inhabitants to stay connected and share their experiences.

  The government stays connected and motivates the residents to give feedback to work more efficiently and effectively. Residents in some smart cities can collect data on air pollution and noise pollution using a low-cost environmental testing kit. The data is streamed to an online platform, where it is compiled into a global crowd-sourced map of data.
- **Reduced environmental footprint:** Littering on the streets, dumping waste in waterways, and increasing greenhouse gas emissions are all bad to the environment. Smart cities are fighting back. Communities now have access to new tools such as energy-efficient buildings, air quality sensors, and renewable energy sources to help them reduce their negative environmental impact.

  Air quality sensors need to be deployed in the entire city, to track the quality of the air and also to identify the cause of pollution and also help analyze the data for further strategic planning (develop action plans) for ensuring the improvement in the quality of air. These sensors can help lay a foundation for reducing air pollution in even the most populated cities, which will undoubtedly save lives.

- **New chances for economic development:** Investments on making smart cities are becoming more significant in strengthening cities' regional and worldwide competitiveness in order to attract new people and businesses. Businesses can also make educated decisions based on data analysis of smart city integrated technologies by providing an open data platform with access to city information.

7.14 Disadvantages of Smart Cities

Smart cities provide numerous benefits for enhancing the safety of their citizens, like smart roadways, surveillance systems, and public safety monitoring, but there are many concerns also to look after. The major concern besides the connectivity of millions of IoT devices/ objects is security. To prevent hacking or misuse, all of this necessitates a solid and secure data collection and storage system.

- **Inadequate privacy:** Maintaining anonymity is challenging due to the deployment of security cameras and intelligent devices connected *through* numerous spaces. To avoid privacy concerns, smart city data must also be anonymized.
- **Social control:** To track and centralize data authorizes the in-charge by giving access to the information, in a short way allowing and empowering the concerned person. Whether it is a government or a private organization, whoever has access to citizens' data has the ability to influence, to scare, and even manipulate public opinion.
- **Excessive network trust:** Cities that rely extensively on networks, electronics, and sensors lose decision-making autonomy and may be unable to react or act if these tools are unavailable or if the network is down.

7.15 Need for Smart City

Several cities are experiencing an overpopulation crisis, resulting in a scarcity of resources. The problems of a community are caused by social and economic imbalances among its citizens. As technology advances, artificial intelligence and the Internet of Things concepts can be applied to the design of smart cities, which can gradually address a variety of issues in a coexisting society [40].

Smart cities are designed to optimally utilize the available space and resources, while also distributing benefits in the most efficient and effective manner possible. It also aims to improve connectivity among citizens and the administration at various levels. Roads, schools, and hospitals are being upgraded as part of the public infrastructure. The system has the potential to eliminate several redundancies in the current system, saving both time and money [40].

The goal of smart cities is to be environmentally friendly. There are devices available that can monitor air purity as well as other environmental and health-related factors. Investment in such a city should also include the maintenance of a conscientious workforce that will review and improve the system. As a result, a smart city will only be considered healthy if it meets the social and psychological needs of its residents [40].

7.16 Smart City Security

Smart cities must be safeguarded against hacking, cyberattacks, and data theft, as well as ensure that the facts reported are correct.

To handle the security of smart cities, resilient authentication management, physical data vaults, and IT solutions must be adopted. Citizens must trust smart city security, which requires coordination between the government, private sector businesses, software developers, device manufacturers, energy providers, and network service managers to create integrated solutions that meet core security objectives. The following are the key security goals.

- **Availability**: In order to perform its duty of monitoring various aspects of the smart city infrastructure, data must be available in real time and with consistent access.
- **Accuracy and accessibility**: The data must be both easy to access and accurate. This includes safeguarding against tampering from the outside.
- **Confidentiality**: Private information must be kept private and protected from unauthorized access. This could entail data anonymization or the deployment of firewalls.
- **Accountability**: Users of sensitive data systems must bear accountability for their actions and interactions. In order to maintain accountability, user logs should document who is viewing the information.

7.17 Smart Cities in the Various Parts of the World

The world is growing smarter by envisaging and utilizing the technology to the best. Cities around the world are at various stages of developing and implementing smart technology. But, some of them are beforehand of the curve and are paving the way to fully smart towns. Following are a few examples.

In the year 2020, London was stated as the smartest metropolis using IESE towns in Motion Index 2020 [7]. London's primary ranking is because of its sturdy overall performance in nearly each measurement: it ranked 1st in human capital and worldwide projection, 2nd in governance and urban making plans, and within the pinnacle ten in mobility and transportation and era. London has more start-ups in step with capital than nearly every other town in the world and has consistently executed nicely on the annual index since 2017.

Singapore is ranked on number 2 in terms of technology, and they are considered as a leading contender in the race to create completely smart cities, with technological advancements in almost all domains. For instance, cameras monitor the crowd density, keep an eye on cleanliness of public places, and also manage the traffic by observing the movement of vehicles. Singapore additionally has real-time tracking systems for electricity use, waste management, and water use. It is also ranked 3rd in phrases of global projection and 7th in terms of environmental effect. Singapore is the weakest in terms of mobility and transportation for which the metropolis is ready to launch driverless taxis and buses by 2022 which will keep them ahead of other towns.

Kansas City is using technology extensively to felicitate its citizens. They have installed interactive kiosks, smart streetlights, and free Wi-Fi. Residents can also view parking space details, traffic flow measurements, and pedestrian hotspots via the city's data visualization app.

Basel is ranked at the 21st position; it is nominated first for social cohesion giving quality life to the citizens, owing to its exceedingly same earnings distribution, low unemployment, less crime, and homicide fees [2, 7].

In the meantime, let us explore San Diego; they have mounted more than 3000 smart sensors to improve visitors glide and parking whilst also increasing public protection and environmental awareness. Solar-to-electric powered charging stations help electric powered cars, and linked cameras reveal visitors and crime.

Dubai additionally has visitors tracking systems in location, as well as smart healthcare solutions, telemedicine and as well as smart buildings, utilities, training, and tourism alternatives.

Barcelona additionally has smart transportation structures, which includes bus stops with free Wi-Fi and USB charging ports, as well as a motorbike-sharing application and a smart parking app with online price alternatives. Sensors that measure temperature, pollution, and noise are also used, as are sensors that measure humidity and rainfall [7].

7.18 Conclusion

Urban issues have reached unprecedented proportions as a result of rising population levels and a sudden population boom in cities. Cities today face significant political, economic, and technological responsibilities that are essential to be met in order to ensure their residents' long-term prosperity. Aside from that, cities face difficulties such as resource scarcity, traffic, pollution, and other things. Smart cities are becoming more common, and governments are growing more reliant on technology to offer a good standard of living for their population. Many cities bring several benefits to inhabitants across the world, not only in terms of enhancing quality of life but also in terms of reducing pollution by offering driverless and electric vehicles and usage of solar energy. It is vital to act swiftly since the necessity to ensure sustainable development and meet the needs of a rising population while minimizing negative environmental impacts will become critical very soon.

It is time to use technology to create intelligent systems to make better use of limited resources. IoT is critical in the development of smart cities because it allows "things" to connect and communicate. Similarly, artificial intelligence is assisting in the analysis and prediction of behavior, allowing users to make informed decisions. Smart cities have a bright future because they improve the standard of living by providing easy access to monitor our health, air, and water quality, traffic management (including parking availability), fleet management, and many other features for providing more accurate and precise information.

For instance, smart parking can help drivers locate a parking space while also allowing them to pay digitally. Another example is smart traffic management, which can optimize traffic lights to reduce congestion by observing the traffic flows. Smart city infrastructure can also manage ride-sharing services.

Smart city characteristics include energy conservation and environmental efficiencies to tackle waste management, climate change, air pollution, and sanitation. Systems like fleet management systems, garbage collection bins that are connected to the internet, and lamps that dim automatically when the roads are empty are used.

Smart cities allow safety measures using sensors by analyzing the data collected from sensors and observing the high-crime areas by providing prior notification about the natural calamities like floods, landslides, storms, droughts, etc. Smart cities need to be resilient instead of focusing on safety and security. Cities need to be prepared to facilitate evacuation emergency reaction processes in the wake of natural calamity or other catastrophic occasions.

Use of AI and deep learning can be crucial in predicting and doubtlessly averting these sorts of activities.

In addition, smart buildings can give real-time structural health monitoring, space management, and feedback to identify when repairs are required. Citizens can also utilize this system to report any difficulties, such as potholes, to officials, and sensors can identify infrastructure issues, such as water pipe leakage. Smart city technology also has the potential to increase efficiency in manufacturing, urban farming, energy use, and other industries.

References

[1] Anonymous (2019). 5 Smart City Technology Trends, ABI Research. Retrieved from: https://www.youtube.com/watch?v=-58A780Ed9o

[2] Anonymous (2019). Why We Need Smart Cities. BWsmartCities. Retrieved from: http://bwsmartcities.businessworld.in/article/Why-We-Need-Smart-Cities/ 19-02-2019-167307/

[3] Anonymous (2020). Advantages and Disadvantages OF SMART CITIES. PrimeStone Intelligent Data Services+ Analytics. Retrieved from: https://primestone.com/en/advantages-and-disadvantages-of-smart-cities/

[4] Anonymous (2021). Internet of Things – IoT Advantages and Disadvantages – 2021. Robu. Retrieved from: https://robu.in/internet-of-things-iot-advantages-and-disadvantages-2021/

[5] Anonymous (n.d). India View. Smart Cities as Envisioned by MoUD. ArcIndia News. Retrieved from: https://www.esri.in/~/media/esri-india/files/pdfs/news/arcindianews/Vol9/smart-cities-envisioned-by-MoUD.pdf

[6] Anonymous (n.d). What are the Disadvantages of AI?. Data Science And Analytics. Pro school. Retrieved from: https://www.proschoolonline.com/blog/what-are-the-disadvantages-of-ai

[7] Anonymous (N.D.). What is a Smart City? – Definition and Examples. TWI. Retrieved from: https://www.twi-global.com/technical-knowledge/faqs/what-is-a-smart-city#SmartCityDefinition

[8] Bansal, S. (N.A). 10 Advantages and Disadvantages of Artificial Intelligence. AnalytiXlabs. Retrieved from: https://www.analytixlabs.co.in/blog/advantages-disadvantages-of-artificial-intelligence/

[9] Chouffani, R. (2020). Future of IoT in healthcare brought into sharp focus. TechTarget. Retrieved from: https://internetofthingsagenda.techtarget.com/feature/Can-we-expect-the-Internet-of-Things-in-healthcare

[10] *Drew, W., (2016).* Internet of Things (IoT): Pros and Cons. Keyinfo. Retrieved from: https://www.keyinfo.com/pros-and-cons-of-the-internet-of-things-iot/

[11] Duggal, N. (2021). Advantages and Disadvantages of Artificial Intelligence. Simplilearn. Retrieved from: https://www.simplilearn.com/advantages-and-disadvantages-of-artificial-intelligence-article

[12] Ersue, M.; Romascanu, D.; Schoenwaelder, J.; Sehgal, A. (2014). "Management of Networks with Constrained Devices: Use Cases". IETF Internet Draft. Retrieved from: http://internetofthingsagenda.techtarget.com/feature/Can-we-expect-the-Internet-of-Things-in-healthcare.

[13] Folk, Chris, Hurley, D.C., Kaplow, D., K., Payne, J., F., X., (2015). Security Implications of the Internet of Things. published in AFCEA International Cyber Committee, Retrieved from: http://www.afcea.org/mission/intel/documents/InternetofThingsFINAL.pdf

[14] Goddard, W. (2019). History of IoT: What It Is, How It Works, Where It's Come from and Where It's Going. IT Chronicles. Retrieved from:

[15] Haas, K. (2021). 5 Ways Smart City Technology Benefits Cities and Residents. Skyfii. Retrieved from: https://skyfii.io/blog/5-ways-smart-city-technology-benefits-cities-and-residents/

[16] Maxwell, J. Clerk A Treatise on Electricity and Magnetism, 3rd ed., vol. 2. Oxford: Clarendon, 1892, pp.68–73.]

[17] Johnson, J. (2020). 4 Types of Artificial Intelligence. Machine Learning & Big Data Blog. Retrieved form: https://www.bmc.com/blogs/artificial-intelligence-types/

[18] Joshi, S., Saxena, S., Godbole, T., Shreya, (2019). Developing Smart Cities: An Integrated Framework, Procedia Computer Science, Volume 93, Pages 902-909, ISSN 1877-0509, https://doi.org/10.1016/j.procs.2016.07.258. Retrieved from: https://www.sciencedirect.com/science/article/pii/S1877050916315022

[19] Jyotsna (2021). 10 Major Applications of IoT You Should Know. Jigsaw. A unext company. Retrieved from: https://www.jigsawacademy.com/top-uses-of-iot/

[20] Kang, Won Min; Moon, Seo Yeon; Park, Jong Hyuk (2017). "An enhanced security framework for home appliances in smart home". Human-centric Computing and Information Sciences. 7 (6). doi:10.1186/s13673-017-0087-4. Retrieved 3 November2017 from:

[21] Kayid, A. (2020). The role of Artificial Intelligence in future technology. Retrieved from: https://www.researchgate.net/

publication/342106972_The_role_of_Artificial_Intelligence_in_future_technology

[22] Khizir, M., Graham, T., Morsalin, E., T., Sayidul, M., Hossain, M.J. (2018). "Integration of electric vehicles and management in the internet of energy". Renewable and Sustainable Energy Reviews. 82: 4179–4203. doi:10.1016 /j.rser.2017.11.004. Retrieved from: https://ideas.repec.org/a/eee/rensus/v82y2018ip3p4179-4203.html y

[23] Khvoynitskaya, S. (2019). The IoT history and future. iTransition. Retrieved from: https://www.itransition.com/blog/iot-history

[24] Kumar, S. (2019). Advantages and Disadvantages of Artificial Intelligence. Towards data science. Retrieved from: https://towardsdatascience.com/advantages-and-disadvantages-of-artificial-intelligence-182a5ef6588c

[25] Lewis, T. (2014). A Brief History of Artificial Intelligence. Live Science. Retrieved from: https://www.livescience.com/49007-history-of-artificial-intelligence.html#:~:text=The%20beginnings%20of%20modern%20AI,%22artificial%20intelligence%22%20was%20coined.

[26] M. Jamshidi et al. (2020). AI and COVID-19: Deep Learning Approaches for Diagnosis and Treatment. Special section on emerging deep learning theories and methods for biomedical engineering IEEE, Retrieved from: https://core.ac.uk/download/pdf/327067595.pdf

[27] Malone, W. T., Rus, D., and Laubacher, R. (2020). Artificial Intelligence and tFuture of Work. Research Brief. Retrieved from: https://workofthefuture.mit.edu/wp-content/uploads/2020/12/2020-Research-Brief-Malone-Rus-Laubacher.pdf

[28] Marr, B. (2018). The Key Definitions of Artificial Intelligence (AI) That Explain Its Importance. Forbes. Retrieved from: https://www.forbes.com/sites/bernardmarr/2018/02/14/the-key-definitions-of-artificial-intelligence-ai-that-explain-its-importance/?sh=7e7ff9314f5d

[29] Merriam-Webster. (n.d.). Artificial intelligence. In Merriam-Webster.com dictionary. Retrieved September 3, 2021, from https://www.merriam-webster.com/dictionary/artificial%20intelligence

[30] *Ramos, J. (2021).* Brief History of Artificial Intelligence. Tomorrow City. Retrieved from: https://tomorrow.city/a/brief-history-of-artificial-intelligence

[31] Rangaiah, M. (2021). History of Artificial Intelligence. Analytic Steps. Retrieved from Blog: https://www.analyticssteps.com/blogs/history-artificial-intelligence-ai

[32] Roznovsky, A. (n.d). 9 Prominent Benefits of IoT for Business. Light of the future. Retrieved from: https://light-it.net/blog/9-prominent-benefits-of-iot-for-business/

[33] Rujan, A. (2018). Thinking about becoming a smart city? 10 benefits of smart cities. Plante Moran. Retrieved from: https://www.plantemoran.com/explore-our-thinking/insight/2018/04/thinking-about-becoming-a-smart-city-10-benefits-of-smart-cities

[34] Russell, S., Norvig, P. (2020). Artificial Intelligence: A Modern Approach, 4th Edition, Pearson

[35] Saxena, P. (2016). The advantages and disadvantages of Internet of Things. E27. Retrieved from: https://e27.co/advantages-disadvantages-internet-things-20160615/

[36] Schroeter, M. (2017). Infographic: Evolution of Artificial Intelligence. Retrievedfrom: https://twitter.com/TCS_Mark/status/861712633518596097

[37] Sharma, M. Aggarwal, N. (2021). Internet of Things Platform for Smart Farming. Internet of Things and Machine Learning in Agriculture. Nova Publications, USA. Retrieved from: https://novapublishers.com/shop/internet-of-things-and-machine-learning-in-agriculture/

[38] Stankovic, J.A. (2014). "Research directions for the Internet of Things," IEEE Internet Things Journal, vol. 1, no. 1, pp. 3–9, Retrieved from: https://www.researchgate.net/publication/264589945_Research_Directions_for_the_Internet_of_Things

[39] Upasana (2020). Real World IoT Applications in Different Domains. Edureka. Retrieved from: https://www.edureka.co/blog/iot-applications/

[40] Vein, C. (2017). Why we need smart cities. PWC. Retrieved from: https://www.pwc.com.au/digitalpulse/why-we-need-smart-cities.html

8

Intelligent Facility Management System for Self-sustainable Homes in Smart Cities: An Integrated Approach

K. Gerard Joe Nigel[1], N. Anand[2], and E. Grace Mary Kanaga[3]

Karunya Institute of Technology and Sciences, India
Email: [1]gerardnigel@karunya.edu; [2]nanand@karunya.edu;
[3]grace@karunya.edu

Abstract

It is expected that 68% of the world's population, over six billion people, will live in megacities and surrounding regions by 2050. A growing population requires the development of an infrastructure and the necessary facilities. Ineffective resource management leads to wasted energy and not meeting sustainability demands in the environment. The utilization of materials for facility management, like water supply and power supplies, do not satisfy the present requirements of sustainability. An increase in the infrastructure of smart cities has raised the demand for integrated facility management systems in residences. It is the responsibility of the individuals to save resources such as energy and water through proper planning and management. Smart materials and intelligent systems play a vital role in improving lifestyles by saving resources without wasting energy. With the help of smart sensor systems and AI-based technologies, it is possible to monitor and control the devices that we use in our day-to-day lives so that facilities and resources can be managed efficiently without wastage. This chapter proposes a framework for an intelligent facility management system for self-sustainable homes in smart cities by integrating building information modeling (BIM), artificial intelligence (AI), Internet of Things (IoT) and big-data analytics (BDA).

8.1 Introduction

Improving the comfort of homes and maintaining their sustainability are the key objectives of a smart home management system. The increasing population in the urban areas creates a need for developing sustainable smart home management systems. It is the expectation that the occupants of a home will enjoy the comfort of living with safety and sustainability. In addition, they need to live a healthy life. Facilities such as electricity, water supply, and safety systems are integrated requirements for such a life. However, the understanding of managing such resources effectively by ensuring their sustainability is very poor.

Buckman et al. [1] defined a smart building system as an infrastructure integrated with intelligence with the aid of suitable sensors to meet the demands of the user by different control strategies and making use of appropriate materials to enhance the energy efficiency of the building system. This can be achieved using appropriate sensor technologies integrated into automation systems. This may improve self-sustainability by saving energy resources with an effective management system. Effective utilization of resources may reduce energy demand.

Smart and sustainable buildings are designed to improve the comfort level of users by changing the environment using self-integrated systems throughout their lifespan. By using smart building systems, the utilization of materials, consumption of water, and energy resources may be optimized to achieve a sustainable and healthy indoor environment [1-3].

The optimum utilization of resources may save energy by improving the circular economy. Advanced sensor systems can effectively monitor and control the required energy. Appropriate automation techniques may be helpful in executing the entire management system, but the system should be economical and durable. A green home can have tremendous benefits, both tangible and intangible. The immediate and the most tangible benefit is a reduction in water and energy consumption. The energy savings could be in the range 15–25% and water savings around 20–50%. Intangible benefits of smart homes include enhanced air quality, excellent daylighting, safe living, lower energy bills, and conservation of natural resources.

The housing sector is growing rapidly in all the cities of India and is contributing to the growth of the circular economy. This creates a need for introducing smart home concepts and intelligent techniques to improve sustainability in resource management. The innovative techniques adopted in the residential sector may be helpful to improve water conservation, energy efficiency, safety systems, and waste management. This may be useful for providing an environment for happy living.

Many of the countries in the Asian region are water-stressed and, for example, in countries like India, the water level underground has significantly reduced over the past decade. The green new buildings rating system (IGBC) highlights the utilization of water in an effective and self-sustainable manner through reducing, recycling, and reusing strategies. By following the guidelines of the rating program, new buildings that are designed as green may reduce potable water consumption by between 30 and 50%.

As a result, building residences in smart cities promotes sustainable urbanization and a healthy lifestyle [4–7]. Smart buildings ensure the well-being of end-users, by satisfying functional needs without affecting the surroundings and nature [2,8–10].

Hence, this study aims to address the significance of different advantages of smart building systems and the key factors responsible for achieving smartness based on the perspective of the user. As smart homes are part of smart cities and their development, this study provides significant insights to be considered by building designers and owners to make living environments more comfortable.

8.2 BIM and Energy Efficient Buildings

Building information modeling (BIM) is not a specific tool, rather it is a technology in which the conceptual stage to the demolishing stage of construction is incorporated. This is not a new technology; it has existed since the 1980s but by different names. BIM is a mandatory tool in growing countries like India. The "I" in BIM, i.e. information, plays an important role. Complete information on a project will be available if the entire work process is modeled using BIM. BIM is a collaborative platform where all models such as architectural, structural, and MEP/HVAC can be shared between the architects, engineers, and construction managers.

A sustainable building is constructed of materials that decrease energy usage. On a theoretical basis, it is very tedious to calculate the energy consumption value. However, by using BIM tools it becomes easy for the construction industry. At the time of the conceptual design stage itself, the solar energy studies can be done. Also, the analysis of the facade system can be done using BIM. In the stage of detailed modeling, the material properties can be defined, so that from the input the energy consumption details can be known using Insight. In the output report, different design options for energy optimization can be included. With the help of the report, the building can be made energy efficient either by changing the material or by using the suitable optimization techniques suggested by Insight.

In the conceptual stage itself, a sun and shadow analysis can be done so that the setting of solar panels and interior design options, etc. can be decided. It is a present requirement that solar paneling be installed in almost every building so that solar energy can be effectively used. In this way, energy consumption can be reduced with the help of BIM tools. The solar study is done based on the location and weather conditions, so the result will be more accurate.

Parametric BIM is an emerging technology in which different design options for facade systems can be adopted for different sun exposure. By including visual programming techniques like Dynamo, different design algorithms can be created by customization. Since it is an algorithm-based concept, just by changing the input the maximum possible outputs for the problem can be generated. This is an advantage of the BIM tool, as it reduces man-hours and optimized cost-effective results can also be achieved.

8.3 BIM in MEP Application

Keeping an eye on industries like the heating ventilating and air conditioning (HVAC) sector, starting from the design stage to operation and maintenance, requires an large labor headcount. The stages involved in mechanical electrical plumbing (MEP) are system design, load estimation, coordination, costing, scheduling, energy simulation, better air quality, effective health environment, and maintenance of facilities with services. These different stages have various expertise involved in carrying out different activities. All these challenges arise due to a lack of coordination between different sectors.

Using BIM from the beginning of a project helps engineers and designers make better decisions earlier in the process. Building information analytics (BIA) also provides analytical data from the design for day-to-day operations. An MEP system provides a complete solution that can design the HVAC system, easily estimate the heat load, costing and estimation, time estimation, visualization, and clash detection.

Three-dimensional smart design software adapting MEP concepts can combine electromechanical systems with building models and provide design and building performance methods for engineers. Simulation results could help the project to understand the energy consumption details by which one can alter the project and achieve energy conservation and maintenance, and operation of the air conditioning system as well. Insertion of MEP BIM information will allow a clearer view of the project as a whole. This leads to a clearer vision throughout the construction phase of the project. BIM in modeling MEP/HVAC applications is extremely useful for integrating customized intelligent buildings.

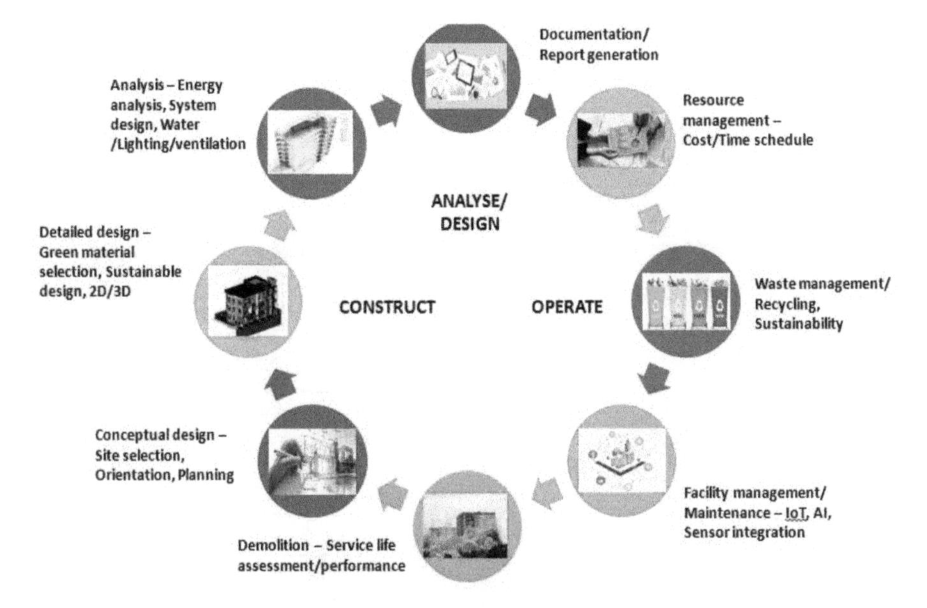

Figure 8.1 BIM process flow.

A building energy management system that reduces the energy consumption and the CO_2 emissions of the building during its operation phase is based on self-learning techniques [11]. Using a BIM process, the integration, design, and operation of a facility management system is easier for the engineers [12]. Figure 8.1 shows the process flow of BIM.

8.4 Integration of BIM and Wireless Sensor Networks (WSNs)

Wireless sensor networks (WSNs) have become popular recently. A WSN is a collection of sensor nodes used for monitoring certain events. The longevity of a WSN depends on the optimal usage of available energy. Optimization is essential for the proper effective utilization of energy. Integrating BIM with a WSN helps improve sustainability.

This technique has been recently adopted and performs the functionalities of 3D CAM/CAD techniques using an object-oriented methodology. Using BIM a virtually developed model of the building is built, which can be used for design, planning, pre-, and post-building processes. The BIM is visible in 3D form and can be applied to other data analytical applications. Complete integration of a WSN and BIM is essential in real-time smart home monitoring systems to stay connected to other systems/environments.

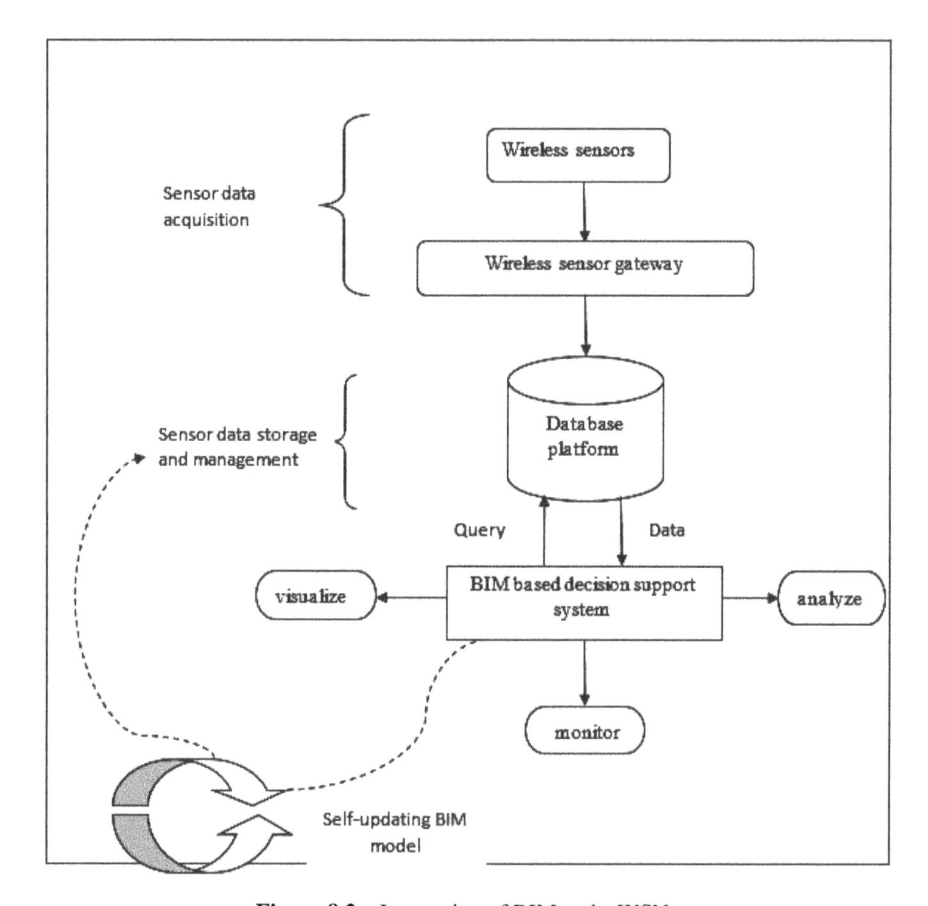

Figure 8.2 Integration of BIM and a WSN.

A three-dimensional model (3D) of a building can be outlined and the sensor wireless devices that are to be deployed to acquire the various parameters of data. A huge amount of real-time sample data is required for this simulation. The processed data can be added to the database using an SQL open-source platform and all sensor stream data registered. The data will be continuously streamed and sent at a predefined interval to the base station. Each application fetches sensor data when needed from the database. Figure 8.2 shows the integration process of BIM and a WSN.

Each sensor device acts as a service provider and has a specific application. Autodesk software architecture is utilized as a platform for modeling the features of buildings and the environment. The model is then exported into an XML file which contains the information on the various employed sensors

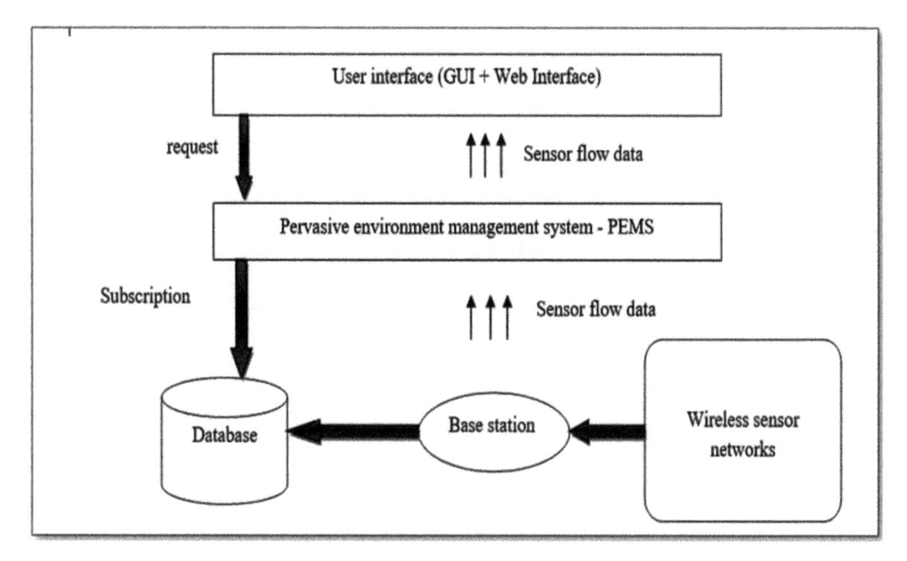

Figure 8.3 The architecture of the proposed system.

and devices. Using Python, an XML parser is developed to read the file and understand the sensor device information. Figure 8.3 displays the architecture of the proposed system.

8.5 AI in Smart Homes

IoT paves the way for communication between various objects and devices embedded with sensors via the internet. Either it can be controlled remotely or the data can be sent to a remote user through AI. The status of a particular device connected to the same network can be monitored and real-time data is obtained utilizing AI.

Incorporating AI into the infrastructure of a smart home is of great help in gathering information from the automation devices, providing maintenance data, predicting user experience, and enhancing data and security. The need for human involvement is limited, with the presence of AI in home automation allowing us to control appliances and devices remotely.

The automating process is mainly based on the information collected by the various devices. It is then trained using an appropriate intelligence algorithm to perform certain tasks automatically. Using the data, it can perform without any human involvement (for example, the Nest thermostat from Nest Labs automatically learns from its user's characteristics and uses the information to set the temperature automatically when no one is at home). Figure 8.4 shows the data exchange levels of powered smart homes.

8.5.1 Data exchange levels of powered smart homes (AI and IoT)

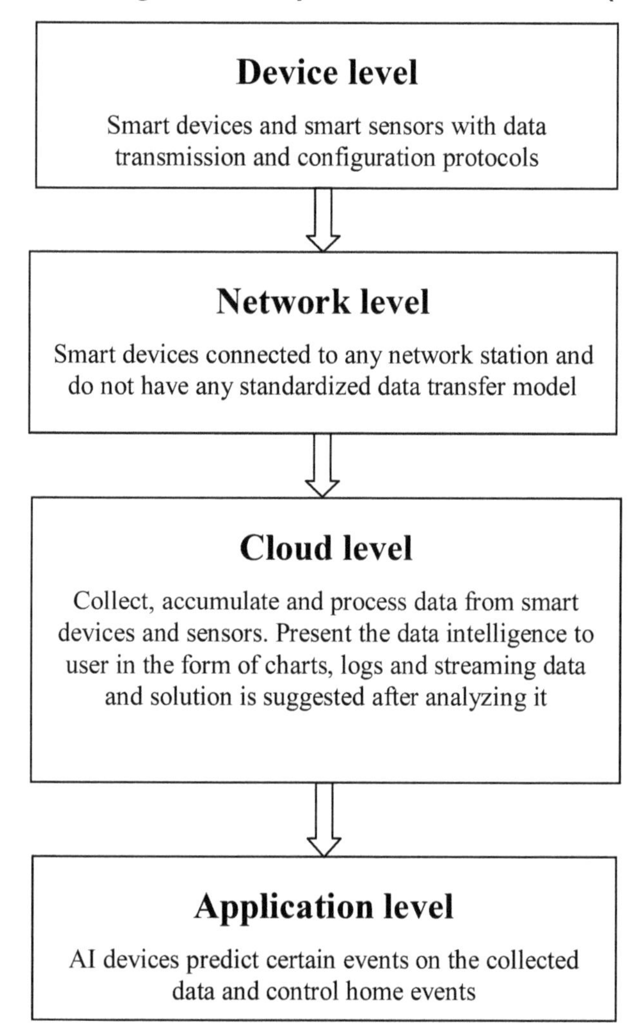

Figure 8.4 Data exchange levels within AI.

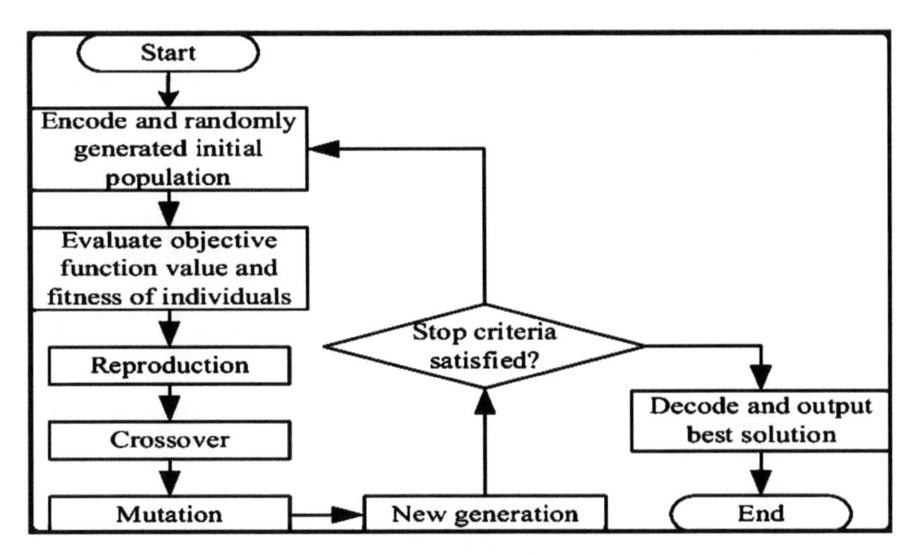

Figure 8.5 GA strategy for energy management of smart appliances in an IoT enabled environment.

8.6 Genetic Algorithm Based Strategy for Efficient Energy Management

A genetic algorithm (GA) can be employed to schedule the smart appliances of a residential building to minimize electricity costs and PAR. The operational parameters and power grid energy are given as inputs to the GA. Using these inputs, the GA estimates the solution for populating the input chromosome. It evaluates the fitness function using a randomly estimated population and the best results during evaluation are stored. It is then passed through the cross-over and mutation phases to obtain an optimal solution. The convergence rate is directly related to the mutation and cross-over probability parameters which act as control parameters of GA. For obtaining optimal power usage scheduling, the cross-over rate is fixed at cr = 0.9 and the mutation rate is fixed at pm = 0.1. The population that is returned after the cross-over and mutation phases will be the optimal population and represents the optimal power usage schedule of household smart appliances. Figure 8.5 shows the process flow of the GA strategy for an efficient energy management system.

8.7 Intelligent Model Algorithms for IoT Application

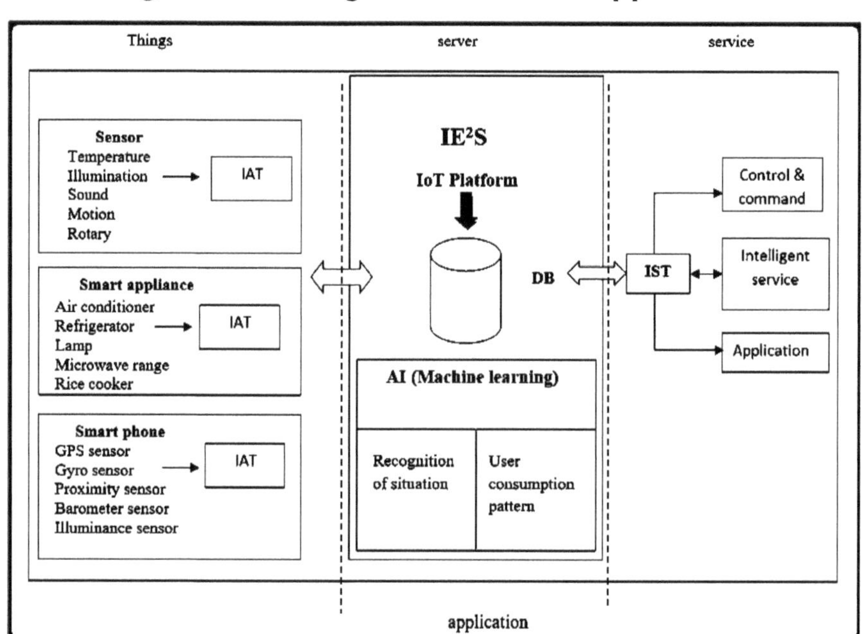

Figure 8.6 Process of an intelligent algorithm using IE2S in an IoT application.

8.8 Intelligence Awareness Target (IAT)

This is the algorithm used to transfer the necessary data to the server from appliances with marginal activity. Only the important information is processed on the server. The various types of data are fed as input. There are three types of data that are considered as input such as sensor data, smartphone data, and smart appliance data in the IAT algorithm. It can be divided into two cases, depending on whether the sensor-based data is entered or smart appliance-based data is entered. The data obtained from a phone or smart appliance is set as the desired prominence data since it helps the user control the device.

Firstly, the data from the sensor is entered. This is done in such a way that various things and sensors operate only when a person is present inside. The filtering process of data is done by activating the filters with the help of sensors, and if this is outside the range the residual data is transferred to the server. When data is sent from a phone or smart appliance, it is considered as user-defined and sent to the server. The data values can be adjusted as the user practices the application of smart appliances, and changes can also be made by them to the anticipated values based on their needs. In this process it may be helpful to know the kind of conditions the user needs to alter the conditions. Figure 8.6 shows the process of incorporating intelligent algorithms using IE2S in an IoT application.

8.9 Intelligence Energy Efficiency (IE2S) Algorithm

IE2S is a server that collects the data obtained from IAT algorithm and it is further processed for learning. It takes the filtered and desired data from the IAT algorithm. To learn from the data, the current status is sent for learning the data from the user's desired data. In case any difference is observed between the present and anticipated status, the IE2S forecasts the optimum value via deep learning. An artificial tensor flow engine is employed for learning purposes. It helps in increasing the accuracy of the forecasted value.

8.9.1 Intelligence service (IST) algorithm

The current status of the devices is maintained or changed using the IST algorithm using the commands received from the IE2S. This algorithm executes the orders obtained from the IE2S and creates a shared communication between devices and the IE2S. It receives commands continually and communicates the status of the device to the IE2S. The data that is transmitted is obtained continually and is analyzed using a neural network algorithm. It helps identify the user consumption pattern. The overall objective is to optimize the energy and utilization of the network, and through this efficiency and automation can be achieved.

Example: The temperature values are arbitrarily considered and the anticipated values are projected by the learning feature of the algorithm. The outcome is identical to the user's preferred value. The blue line indicates the average temperature (monthly data). The black line denotes the value of temperature preferred by the user. The red line displays the ultimate estimated value through learning. The red and the black lines are nearer to the blue line and when more information is gained, errors may be eliminated in the further process. Figure 8.7 shows the sample estimation of temperature through intelligence learning.

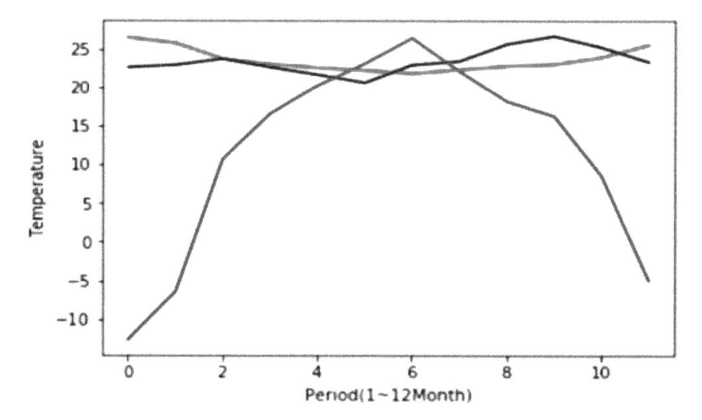

Figure 8.7 Estimation of temperature through intelligence learning.

8.10 BIM and Big Data Analytics

IoT, big data, and machine learning (ML) are the emerging tools widely used for different applications. These technologies are set to transform all areas of business as well as everyday life. ML takes a large amount of data and generates useful insights. Most of the class data models are often static and of limited scalability and hence not suitable for fast-changing and fast-growing volumes of unstructured data. The technology that supports the mentioned challenges is big data analytics. In this section, the need for big data in BIM and its analytical model are discussed. The integration of a WSN in BIM generates a huge amount of data. This IoT-generated data can be used for real-time decision making as well as for doing offline analysis of historical data which could be used for designing smart intelligent buildings [13]. Big data analytics plays a vital role in processing and analyzing this data. The analytical model based on big data technologies for BIM is shown in Figure 8.8.

Based on the requirements, the intelligence of our system is provided by the combination of IoT devices and big data analytics. Various sensors are used to measure energy and water usage, and to monitor the security of the smart building. These sensors send the data over the wireless networks to the processing server, where big data analytics needs to be employed to analyze the data and to produce useful insights and alerts [14]. Predictions regarding BIM can also be done via analytics. Each module of the analytical model based on big data technologies for BIM is described as follows:

8.10.1 Data acquisition

Data collection is a major part of building the BIM model. Training the model without sufficient data will prevent the model from generating meaningful

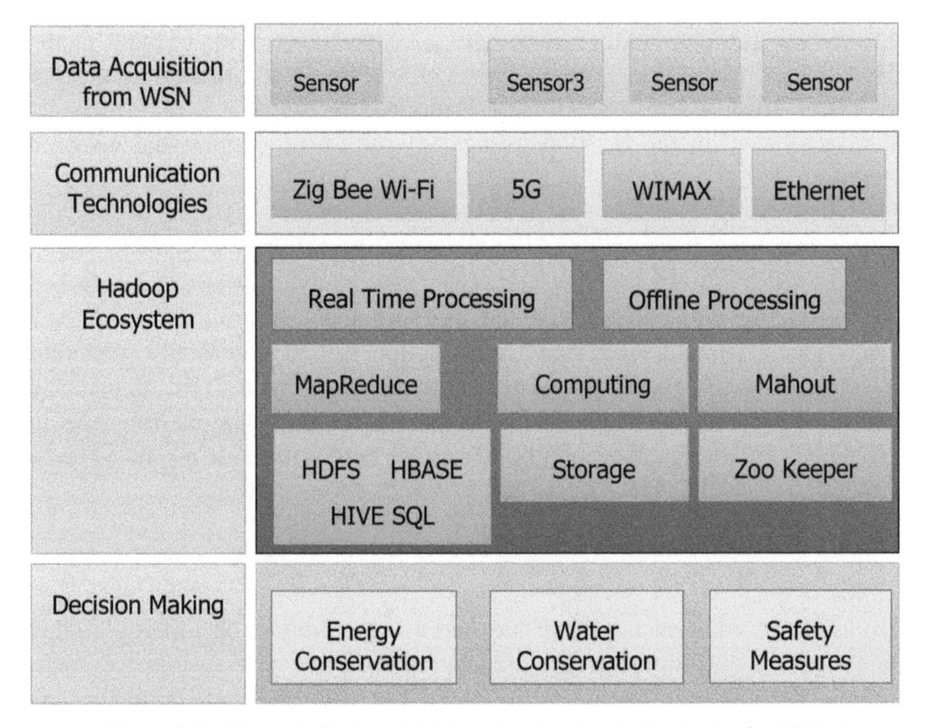

Figure 8.8 The analytical model is based on big data technologies for BIM.

outcomes. Data acquisition is done by deploying various sensors in the smart building. The data covers a variety of information from the building like water consumption, energy consumption, and other security-related parameters that need to be monitored.

8.10.2 Communication technologies

The data generated by the various sensors in the WSN is sent to the central processing system via a high-speed wireless transmission medium. The widely used communication technologies are 5G, Ethernet, WiFi, and WiMax.

8.10.3 Hadoop ecosystem

The Hadoop Ecosystem helps in the implementation of data management and processing systems for BIM. It receives the data generated by the various sensors that are deployed and processes them according to the specified algorithm. MapReduce and HDFS are the core components of the Hadoop framework for computation and storage respectively [15]. Mahout is the Hadoop ecosystem that has built-in machine learning algorithms. Zookeeper monitors all other processes of the Hadoop ecosystem. The pre-processed data are stored in the Hadoop storage for further processing and decision making.

8.10.4 Decision making

Processing the data and applying the appropriate decision support model is a major part of this model. Energy consumption will be monitored in different scenarios and, based on historical data analysis, decisions will be taken to reduce energy consumption. Energy conservation is considered important in BIM [16]. IoT devices in smart buildings are used to provide advanced technology to control smart systems and to reduce energy wastage.

8.11 Conclusion

Building information modeling has had a major impact on urban planning and national development. In this chapter, we propose a system for smart intelligent buildings by using IoT-based big data analysis. It touches on functionalities, data collection and aggregation, data cleaning, communication, processing, interpretation, and decision making. The main challenge is the deployment of sensors in appropriate locations of the smart building. Therefore, with the aid of modern learning tools, it is possible to integrate facility management systems for improving the self-sustainability of smart homes.

References

[1] A.H. Buckman, M. Mayfield, S.B.M. Beck, 'What is a smart building? Smart Sustain', Built Environ., 2014, 3, 92–109.

[2] N. Attoue, I. Shahrour, R. Younes, 'Smart building: Use of the artificial neural network approach for indoor temperature forecasting', Energies, 2018, 11, 395.

[3] W.M. To, P.K.C. Lee, K.H. Lam, 'Building professionals' intention to use smart and sustainable building technologies—An empirical study', PLoS ONE, 2018, 13, e0201625.

[4] P. Siano, I. Shahrour, S. Vergura, 'Introducing Smart Cities: A trans disciplinary journal on the science and technology of smart cities', Smart Cities, 2018, 1, 1–3.

[5] Z. Allam, P. Newman, 'Redefining the smart city: Culture, metabolism and governance', Smart Cities 2018, 1, 4–25.
[6] M. Deakin, A. Reid, 'Smart cities: Under-gridding the sustainability of city-districts as energy efficient-low carbon zones', J. Clean. Prod., 2018, 173, 39–48.
[7] X. Zhang, D. Hes, Y. Wu, W. Hafkamp, W. Lu, B. Bayulken, H. Schnitzer, F. Li, 'Catalyzing sustainable urban transformations towards smarter, healthier cities through urban ecological infrastructure, regenerative development, eco-towns and regional prosperity', J. Clean. Prod., 2016, 122, 2–4.
[8] D. Clements-Croome, 'Sustainable intelligent buildings for people: A review', Intell. Build. Int., 2011, 3, 67–86.
[9] M.V. Moreno, M.A. Zamora, A.F. Skarmeta, 'User-centric smart buildings for energy sustainable smart cities', Trans. Emerg. Telecommun. Technol., 2014, 25, 41–55.
[10] D. Arditi, G. Mangano, A. De Marco, 'Assessing the smartness of buildings', Facilities, 2015, 33, 553–572.
[11] IGBC Green New Buildings Rating System (Version 3.0 with Addendum 5) September 2016.
[12] Autodesk BIM - An overview of BIM, Intelligent 3D modeling for architecture, engineering, and construction professionals.
[13] B. Li, & J. Yu, 'Research and Application on the Smart Home Based on Component Technologies and Internet of Things', Procedia Engineering, 2011, 15, 2087–2092.
[14] A. Daissaoui, A. Boulmakoul, L. Karim, & A. Lbath, (). 'IoT and Big Data Analytics for Smart Buildings: A Survey', Procedia Computer Science, 2020, 170, 161–168.
[15] A. T. Hashem, V. Chang, N. B. Anuar, K. Adewole, I. Yaqoob, A. Gani, … H. Chiroma, 'The role of big data in smart city, International Journal of Information Management, 2016, 36(5), 748–758.
[16] M. Alaa, A. A. Zaidan, B. B. Zaidan, M. Talal, & M. L. M. Kiah, 'A review of smart home applications based on Internet of Things, Journal of Network and Computer Applications, 2017, 97, 48–65.

9

Artificial Intelligence and IoT for Smart Cities

Snehal A. Bhosale

RMD Sinhgad School of Engineering, India
Email: snehal.a.bhosale@gmail.com

Abstract

According to United Nations predictions by 2050, the world's population will reach 9.7 billion, with 70% living in urban areas. Resources like energy, water supply, drainage, and traffic management, as well as associated government services like sanitation, sewerage, and traffic management must be provided to all habitats at the same time without causing environmental degradation, which is challenging. AI-based IoT is the solution to getting rid of these problems. The Internet of Things has become increasingly popular in the last couple of decades due to advancements in hardware and software technology. Hardware advancement has seen a huge increase in functionality with a size reduction. The number of Internet of Things (IoT) devices worldwide is expected to triple from 8.74 billion in 2020 to more than 25.4 billion IoT devices by 2030.

With these developments in hardware and wireless technologies, small objects or things are connected to the internet to form IoT networks. Artificial intelligence (AI) empowers IoT networks by making them take calculative decisions without human involvement. With AI-enabled IoT, one can take advantage of the benefits and convenience of advancements in technology to make everyday life more convenient and secure. With smart technology, smart cities can provide quality and performance services like traffic management and police surveillance, smart grid, smart waste management, and smart governance at reduced costs and high speeds. There are many IoT applications for smart cities that utilize both machine learning and deep learning. This chapter describes those applications. Smart city applications today face many challenges, which are addressed as well.

9.1 Introduction

There is a prediction that over 10% of the population of urban areas will increase in the next 10 years due to an increase in population in the coming years [1]. With the overall population growth, over the next three decades, urbanization will add another 2.5 billion people to the city. Sustainable economic, social, and environmental policies are essential to keep up with this rapid expansion of our city. To provide quality of life for all citizens with proper resources and without an increase in the cost of living is a big challenge for the government system. In this view, smart city has come with solutions along with advanced technologies, information and communication technology (ICT), and various physical devices connected to the Internet of Things (IoT) network to optimize the efficiency of city operations and services and connect to citizens [2–5]. The definition of a smart city manages its resources efficiently. Traffic, public services, and disaster response should be operated intelligently to minimize costs, reduce emissions, and increase performance.

IoT is the system that integrates different technologies through various devices in such a way that the least human intervention is required. This property of IoT enables its popularity in smart city deployment, for better comfort, sustainable living, and productivity of citizens. In this chapter, we are discussing various domains related to smart city implementation using IoT and artificial intelligence (AI). We will discuss the fundamental components, enabling technologies, applications, and challenges concerning IoT-based smart city [6].

In a smart city, ICT is used for the betterment of the livelihood of people living there, and the municipality of that city uses ICT facilities to improve operational efficiency and improve government services by sharing information with the public. An AI-based smart city plays a critical role in ensuring that existing resources and infrastructure are utilized in an efficient and effective manner. With the use of smart technologies and data analytics, improvement of quality of life for its citizens, optimization of city function, and drive economic growth are expected in the smart city. The technologies used in smart city formation are IoT, AI and cloud-based services. The data from IoT-based devices such as connected sensors, meters; lights, etc., are collected and analyzed to improve service, public utilities, and infrastructure. Figure 9.1 gives an idea about various applications of IoT-enabled smart city [7, 8].

During the past few years, internet usage has grown dramatically, making access much easier. Also, the cost of connection is decreased. A lot

Figure 9.1 Smart city applications with IoT and AI.

of new devices have come out with built-in Wi-Fi capabilities and sensors in recent years. A variety of devices are available with a low power consumption, internet connectivity, and a smaller form factor. As a result of these factors, IoT has become one of the most popular technologies today. Nowadays, internet-enabled devices are becoming more common since they are able to communicate with one another and with other internet-connected devices. These devices or things are objects, vehicles, clothing, or environments. A good example of such smart networking through the internet is smart parking. In a smart parking app, a user can easily find parking space availability by avoiding the long queue and saving time and petrol. Other examples are smoke detectors, cargo vehicle fleets technology, farming smart networking, etc. [9].

9.1.1 Impact of AI in smart city

AI is an advanced technology that makes machines act on their own without human intervention using computer programs, where the computer is made to understand human intelligence. The smart city consists of many sensors and actuators, in many applications like traffic sensing, video cameras, environment sensing, vehicles, mobile phones, and smart meters. These sensors and actuators generate huge data which is analyzed; a useful pattern will be extracted from all these sensors, using AI techniques.

9.1.2 Advantages of smart city

i. **Money value:** Nowadays, smart cities are receiving substantial investment. The smart city sector is expected to generate more than 1 trillion dollars by 2025. The government, as well as the private sector, will benefit from this huge investment. Using advanced technologies will provide them with good revenue and incentives in development [10].

ii. **Climate goal achievement:** Among the smart features of smart cities are smart transportation, smart energy management, and smart administration. To reduce carbon emissions and live cleaner and healthier, this plan aims to reduce emissions. Smart cities use the latest technology to improve the environment and have a sustainable climate [11].

iii. **Societal performance:** The development of smart cities involves the use of ICT, which will benefit people living in these cities by improving their quality of life. In a smart city, the foundations are built upon societies; so most domains can be achieved by developing them [6].

9.2 Implementation of Smart City

By observing the components that make up the smart city, it becomes obvious that their development starts with the city itself. These include smart homes, smart energy, smart infrastructure, smart transport, smart health, and smart agriculture. The development of an IoT architecture for the smart city begins when IoT is integrated into the smart city implementation. To develop an IoT-based smart city, a variety of challenges must be resolved, including the design of smart sensors, networking issues, security and privacy issues, and the data analytics of a vast database. Prior to final implementation, sensing technologies must be assessed and studied. It is challenging to develop a smart city if you do not have a good understanding of network topologies, networking architecture, and network protocols. A smart city team must select the appropriate topology, architecture, and protocols based on the application. AI is a key component to success when implementing IoT-based smart cities. By combining machine learning (ML) and deep learning (DL), AI can make smart city implementation an intelligent solution that improves the quality of life of the city's residents.

9.2.1 Aspects of smart city

The process of creating a smart city involves four aspects: a collection of the data, exchange of the data, storage of the data, and the analysis of the stored data.

Data collection, or the first aspect of smart city operation, is the most important aspect. A smart city's sensors collect data from a number of applications that run on them. It is important to gather these sensors data for future analysis. In the next stage, data will be transmitted and received, where data collection points will exchange data and then data will be uploaded to the cloud for storage and analysis. A Wi-Fi network and 4G and 5G networks are used to accomplish this. The exchange of this data takes place both locally and globally through different types of networks.

The third stage in the development of a smart city is the storage of the collected data. Depending on the storage scheme, the data is either stored on-premises or in the cloud. It has been categorized using storage schemes so that it can be analyzed in the next step. Smart city implementation involves analyzing data as one of the most crucial phases. The extraction of patterns is carried out at this stage to ensure proper decision-making. Data analysis is carried out using aggregation for simple data types and advanced methods like ML and DL for complex types of data [12].

9.2.2 Smart city components

Different smart city components will be discussed in this section as shown in Figure 9.2. The first component of the smart city is a **smart home**. It consists of various sensing units installed in the person's house which will provide real-time information about the person's house as well as surrounding conditions. Examples of sensors used in smart homes are motion sensors, gas sensors, timers along with the sensors that will measure power and energy consumption, and web cameras.

Smart city services are next on the list of smart city components. These services are useful for city dwellers in their daily lives. It mainly involves municipal tasks such as water supply management, environment control monitoring, waste management, and quality of life. In water supply management, sensors are used to check the quality of the water, water leakage through pipes, etc. [13]. The sensors are also used to maintain and monitor the environmental conditions, by checking the pollution level of air such as checking carbon dioxide (CO_2) levels in the surrounding air. Waste management is also a very important aspect of smart city services. Sensors can be

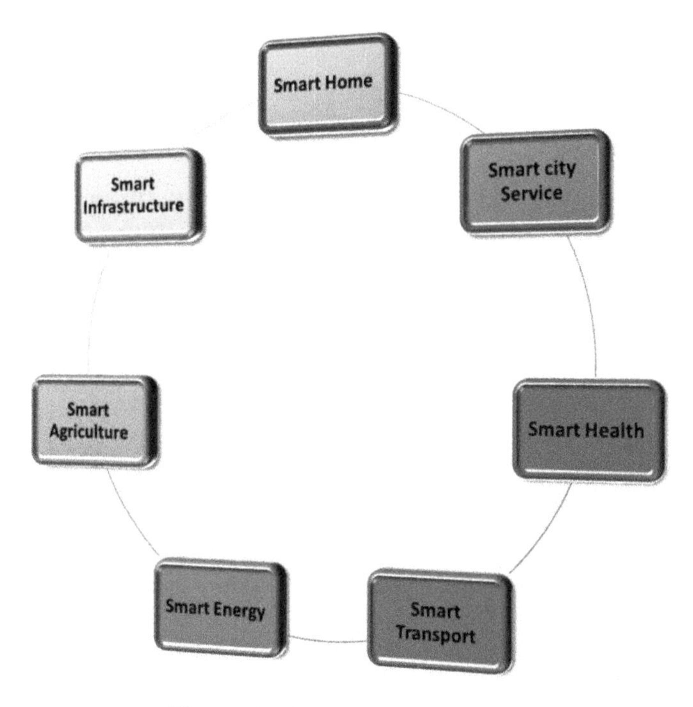

Figure 9.2 Smart city components.

used to segregate the dry and moist waste. The sensors will monitor the level of dust-bins and inform to pick those via the cloud and AI. Smart city services also provide smart parking [14, 15].

Integrated **smart grids** and ICT facilities for smart energy management are the next aspects of smart cities. It makes grids more observable, which allows fault detection and repairs to be done automatically; it also allows energy to be distributed to consumers and utilities simultaneously. Using smart grids, consumers can monitor their power usage in real time, which is then used to generate a power supply, based on predictive models. To ensure the citizens of the city have a continuous power supply, these models were developed using the utilized power data, integrating different energy sources, and using self-healing techniques in the network [16].

In **smart infrastructure**, bridges, roads, and buildings can be constructed with the proper planning so that the facilities of the resources can be used in an effective, uninterrupted manner. Smart sensors such as accelerometers are used to monitor the structural integrity of old buildings, bridges, and heritage structures. By using the data from these sensors, predictive maintenance can be performed so that cities can operate more efficiently [17].

Smart transport is one of the very important elements of smart city implementation. Especially in urban areas, traffic problems such as traffic congestion, parking issues, and pollution are major problems that must be dealt with on a priority basis. Smart transport has a solution for it, where, with the ICT facilities, it offers solutions such as toll systems, vehicle tracking, traffic management, smart parking, accident prevention, etc. Using a technology-enabled mobile app, individuals can navigate the traffic patterns and track their cars for a more convenient trip. GPS-enabled cars can be tracked and observed in real time [18].

Increasing productivity and reducing costs are major issues in the industrial world. **Smart industry**, also known as Industry 4.0, industrial IoT (IIoT), or smart manufacturing, offers a solution. 40% of the world's electricity is used by factories. IoT will reduce factory power consumption in the smart industry by improving its safety, efficiency, flexibility, and environmental impact. Smart factories will be simpler, safer, more efficient, and more environmentally friendly. In factories, machines are connected to the internet and also to the cloud for analysis of the manufacturing to improve the outcomes. However, the connectivity between heterogeneous devices and machines poses many challenges for IoT-enabled industries. As part of industrial 4.0, a cyber–physical system is required for maintaining flexibility and connectivity in these devices. AI is integrated for predicting maintenance, monitoring, and managing production [19–21].

In the context of **IoT-enabled healthcare**, ICT facilities are used to improve the quality and accessibility of healthcare. As a result of the systematic and widespread use of technologies such as wearable health monitors, the creation of open data platforms for health parameters, and the development of virtual communication between patients and healthcare professionals, smart cities can serve as a global health information base [22, 23].

9.3 Artificial Intelligence (AI) in Smart City

9.3.1 AI-enabled smart city applications

Data analytics is an important stage in getting the best results for the inhabitants of the smart city. A smart city gathers data from all sensors and actuators and analyzes it to find hidden patterns that can be used to guide planning and policymaking for a better life. This data is stored in a cloud and analyzed for plans. Using ML and DL algorithms, smart city analytics, which use data analytics to predict the future, can produce better results. Algorithms employed in AI generate automatic results based on data analytics predictions [24, 25].

9.3.1.1 Machine learning

Machine learning (ML) algorithms with data analytics will be very helpful in improving the overall performance of smart city services. AI with ML combined with analyzed data of the smart city can address any type of issue related to healthcare, energy, traffic, agriculture, etc. The huge data generated by sensors and actuators of the smart city will be computed using intelligent methodologies. AI and ML will solve large-scale complex problems in smart city through mathematical computation. The most commonly used ML algorithms for smart city are decision tree (DT), logistic regression (LR), support vector machine (SVM), random forest (RF), naïve Bayes (NB), and *K*-nearest neighbor (*K*-NN) [25].

9.3.1.2 Deep learning

Deep learning (DL) is a ML technique that is used effectively to gain insights from the data, understand the patterns from the data, and predict the data. DL algorithms process noisy data so that output for classification and prediction tasks is provided. DL is known for its ability to handle large volumes of relatively messy data, including errors in labels and large numbers of input variables. In an IoT-enabled smart city, data is generated by heterogeneous sensors which are varying in nature. DL will perform both feature extraction and classification/prediction for this data. For smart city applications, the most commonly used DL algorithms are deep neural network (DNN), convolutional neural network (CNN), recurrent neural network (RNN), stacked auto-encoder network (SAE), and gated recurrent unit (GRU) [25].

9.3.2 Applications of AI in smart city applications

i. **AI in smart home:** For the smart home, ML and DL are used in monitoring applications like activity recognition, fall detection, occupancy detection, localization of a person, automation in energy, and management to reduce power consumption. AI algorithms used in smart home applications are RNN [26], SAE [27], RNN [28], DNN [29], NB [30], and SVM [31].

ii. **Smart city services:** The applications like air quality monitoring, water quality monitoring, waste management, and urban noise monitoring are implemented, analyzed, and monitored with ML and DL. The AI algorithms used in these applications are KNN SVM [32], DNN [33, 34], and RNN [35].

iii. **Smart energy:** With AI, for smart energy applications, energy and load consumption forecasting are done using *K*-means [36], KNN [36], SVM [37], RNN [38], and DNN [39] algorithms. Smart and line event classification is made with DT [40] algorithm. Further, AI is used for electricity theft detection with CNN [41] algorithm.

iv. **Smart infrastructure:** Under smart infrastructure, structural health monitoring is maintained with *K*-means [42], DNN [43], CNN + RNN [44], and KNN [45] AI algorithms. Further, to have better energy and environmental management, SVM [46], SAE [47], and CNN [48] algorithms are used.

v. **Smart transport:** In smart transport, applications like smart parking, occupancy detection, and location prediction are implemented with various AI algorithms. Those are: *K*-NN [49], *K*-means [50], RNN [51], and DNN + CNN [52]. Public transport is implemented using the *K*-means algorithm [18, 53]. Traffic flow and accident detection are managed using RNN [54] and SAE [55] AI algorithms.

vi. **Smart health:** In smart health application, human activity recognition, fall detection is done using DT [56], CNN [57], and SVM [58] AI algorithms. Patient health monitoring is implemented with SVM [59], RNN [60], and CNN [61] AI algorithms. Further, disease diagnosis is performed with DT [62], *K*-means [63], RF [64], and DNN [65]. Parkinson detection is done with RF [66] and seizure monitoring is done with SVM [67], *K*-NN [68], and NB [69] AI algorithms.

vii. **Smart agriculture:** Under smart agriculture, crop monitoring, plant care, disease detection, and irrigation monitoring are performed with LR [70], DT [71], CNN [72], SVR + *K*-means [73], SVM [74–78], and CNN [79] AI algorithms. Predicting physical parameters for decision-making for crop growth is achieved by CNN + RNN [80], RF [81], and DNN [82, 83] AI algorithms.

9.4 Challenges in IoT-based Smart City Implementation

There are the following challenges in IoT-enabled smart city: privacy and security, networking issues, smart sensors, data analytics, cloud computing, and fog computing as shown in Figure 9.3. These challenges are discussed briefly in the following section.

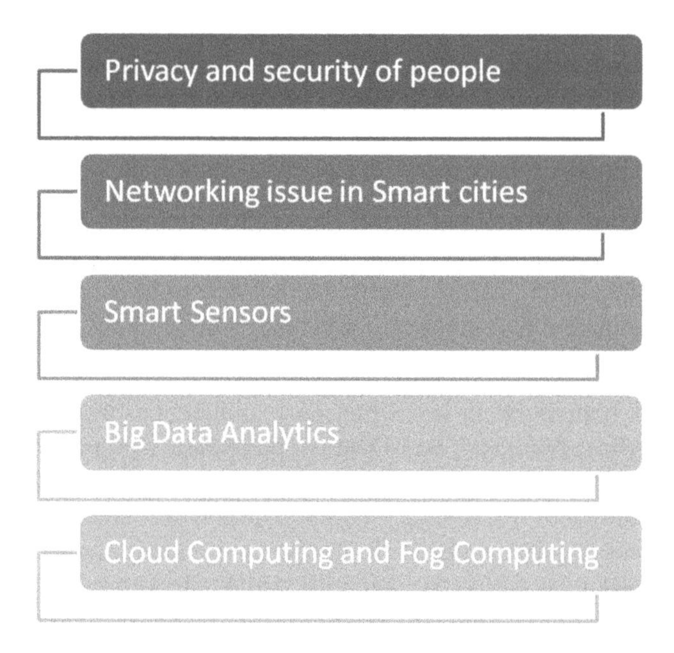

Figure 9.3 Challenges in smart city.

i. **Privacy and security of people:**
Privacy and security are a big concern of IoT-enabled smart city. People using sensors and internet-driven smart appliances may be hacked and the personal data of an individual like username–password, daily routine, medical data, or financial data information may get leaked and misused by intruders for their benefit. This is because smart devices all share information through the internet, which is insecure, and it is simple for hackers to steal someone's information through such a service. Hacking such information could negatively impact citizens of smart cities and threaten their privacy as well as safety. Therefore, privacy and security issues need to be addressed from a priority standpoint. Using encryption to transmit data will reduce the likelihood of cyber hackers gaining access to it. The IoT network's resource-constrained nature makes it difficult to use the cyber-security algorithms of the computer network. Researchers are still working on finding the best security and privacy solution for smart city residents, financial sectors, and industries [84].

ii. **Networking issue in smart city:**
All the information in a smart city is communicated perfectly between different devices as well as to and from the cloud if the underlying

network functions properly. As IoT networking is concerned, there are a lot of challenges in functionality such as bandwidth issues, power consumption, latency, quality of service, communication via heterogeneous protocols, wired and wireless connectivity reliability, communication of encryption, and decryption of data. The current network system is not well upgraded to have smooth communication between all smart devices.

Many devices in the smart city are in moving/mobile condition. A smart city application requires making sure that the quality of the service is maintained without interrupting the connection. Networking at the local level as well as globally is equally important for providing a proper response. The routing protocols used for global communication are important since they should ensure the best quality and the lowest possible delay in sending data from one device to another. Security is one of the key considerations for the transmission of data from one smart device to another. Several protocols do not provide the network services perfectly necessary for a smart city to operate smoothly. All these issues should be resolved with a formal solution [85].

iii. **Smart sensors:**

Sensors are the heart of the IoT network. They sense the physical phenomenon and convert it into an electric signal. However, these sensors, for example, temperature sensors, moisture sensors, pollution sensors, IR sensors, etc., are manufactured by different vendors that work with different protocols. In a smart city, these sensors need to gather various data and pass within different sensors networks, to the cloud and from the cloud. All vendors and manufacturers can work together to solve the heterogeneous sensor problem in order to maintain a common platform for sensor communication and facilitate data exchange. Sensors in IoT are also required to consume minimal power. These smart sensors must be designed so that they consume as little energy as possible by the manufacturers of sensors and smart city designers. To ensure the same output for similar conditions in the long run, sensors must be reliable and robust. The quality of the service should be maintained throughout the lifetime of the sensors.

Data that is collected by smart city sensors need to be stored by these sensors for many applications. Thus, manufacturers of sensors should manufacture robust sensors that have storing capability and compressed algorithms which will consume less power for processing the data generated in different applications of the smart city [16, 86].

iv. **Big data analytics:**
In an IoT-enabled smart city, a huge amount of structured and unstructured data is generated. For a smart city to function effectively, the collected data from smart sensors needs to be analyzed so that specific patterns are created and forecasted using AI techniques. In big data analytics, new algorithms need to be developed to improve the quality of service in smart city applications. As a variety of data will be generated in smart city applications, these algorithms must accommodate all types of videos, audios, images, texts, social media, blogs, financial transactions, smart health records, and satellite imagery. In order to expand the smart city services, resources, or areas, this variety of data should be analyzed properly to make wise decisions for the sake of the smart city inhabitants. ML and DL are used in advanced smart-city-based big data analytics for better decision making [7, 87, 88].

v. **Cloud computing and fog computing:**
Cloud computing provides data storage and scalable processing power for various IoT applications of a smart city. Different sensors of smart city applications are connected to cloud computing services of the smart city to collect, process, and store this data. Cloud computing provides the necessary platforms to store and process the sensor data for the operation and planning enhancement of smart city. For smooth functioning of the sensors and actuators data with cloud computing, the strong network architecture of the smart city is required. The services provided by cloud computing are either offered at a centralized location or distributed platform at various locations. Due to benefits like better quality and reliability support, distributed cloud computing is preferred. However, good communication links need to be available in a different location for the smooth functioning of distributed cloud computing. If communication links are not consistent, then issues like transmission delay, loss of packets, and unstable connections will take place which will affect the smart city performances [89–91].

Many issues are not solved by cloud computing alone; hence, we need to use fog computing for beneficial services. Those issues are mobility, real-time data generation and processing, lossy links, synchronization, low latency, etc. These issues happen because of the distance between sensors and actuators connected to smart devices and cloud computing platforms. Whereas, fog computing offers more localized services that offer fewer delays. With this approach, IoT applications of the smart city become more localized, mobile, and fast-responding. However, without good networking support, fog computing will not be successful [89, 90].

9.5 Conclusion

IoT-enabled smart cities that incorporate AI as a vital component of the IoT infrastructure were discussed in this chapter. A smart city is an urban area that uses information and communication technologies to improve the quality of life for its residents. Smart cities rely on technology such as artificial intelligence (AI) to make effective and efficient use of their infrastructure. Using data analytics and smart technologies, a smart city aims to enhance its occupants' quality of life while spreading economic growth. The science of AI is the creation of intelligent computer programs that can learn about human behavior and intelligence. Often, AI is used to gather and analyze a large amount of data. Predictions based on the hidden pattern of analyzed data are made for developing more intelligent solutions for enhancing the lives of smart city citizens. Several components of smart cities are presented here. There were different algorithms discussed for the implementation of smart city applications using ML and DL. Many challenges need to be addressed in smart city implementation in order to ensure success, and several aspects of smart city implementation will be beneficial. Since cities' populations are growing at a fast pace, smart city implementation is ideal for making a city a better place to live. As a result of advanced technologies such as IoT, data analytics, and AI, the implementation of a smart city appears not to be difficult.

References

[1] Worldometers. World Population Forecast—Worldometers. 2019. Avaliable online: https://www.worldometers.info/world population/ world-population-projections/ (accessed on 11 July 2021).

[2] Cardullo, P.; Kitchin, R. Being a 'citizen' in the smart city: Up and down the scaffold of smart citizen participation in Dublin, Ireland. GeoJournal 2019, 84.

[3] Desdemoustier, J.; Crutzen, N.; Giffinger, R. Municipalities' understanding of the Smart City concept: An exploratory analysis in Belgium. Technol. Forecast. Soc. Chang. 2019, 142, 129–141.

[4] Khan, M.S.; Woo, M.; Nam, K.; Chathoth, P.K. Smart city and smart tourism: A case of Dubai. Sustainability 2017, 9, 2279.

[5] Wu, S.M.; Chen, T.C.; Wu, Y.J.; Lytras, M. Smart city in Taiwan: A perspective on big data applications. Sustainability 2018, 10, 106.

[6] Hollands, R.G. Will the real smart city please stand up? Intelligent, progressive or entrepreneurial? City 2008, 12, 303–320.
[7] Sánchez-Corcuera, R.; Nuñez-Marcos, A.; Sesma-Solance, J.; Bilbao-Jayo, A.; Mulero, R.; Zulaika, U.; Azkune, G.; Almeida, A. Smart city survey: Technologies, application domains and challenges for the city of the future. Int. J. Distrib. Sens. Netw. 2019, 15.
[8] Silva, B.N.; Khan, M.; Han, K. Towards sustainable smart city: A review of trends, architectures, components, and open challenges in smart city. Sustain. City Soc. 2018, 38, 697–713.
[9] Ejaz, W.; Anpalagan, A. Internet of things for smart city: Overview and key challenges. Internet Things Smart City 2019, 1–15.
[10] Anthopoulos, L.G.; Reddick, C.G. Understanding electronic government research and smart city: A framework and empirical evidence. Inf. Polity 2016, 21, 99–117.
[11] Atat, R.; Liu, L.; Wu, J.; Li, G.; Ye, C.; Yang, Y. Big data meet cyber-physical systems: A panoramic survey. IEEE Access 2018, 6, 73603–73636.
[12] Khan, Z.; Anjum, A.; Soomro, K.; Tahir, M.A. Towards cloud based big data analytics for smart future city. J. Cloud Comput. 2015, 4.
[13] Rojek, I.; Studzinski, J. Detection and localization of water leaks in water nets supported by an ICT system with artificial intelligence methods as away forward for smart city. Sustainability 2019, 11, 518.
[14] Dutta, J.; Chowdhury, C.; Roy, S.; Middya, A.I.; Gazi, F. Towards smart city: Sensing air quality in city based on opportunistic crowd-sensing. ACM Int. Conf. Proc. Ser. 2017.
[15] Al-Turjman, F.; Malekloo, A. Smart parking in IoT-enabled city: A survey. Sustain. City Soc. 2019, 49.
[16] Shirazi, E.; Jadid, S. Autonomous Self-healing in Smart Distribution Grids Using Multi Agent Systems. IEEE Trans. Ind. Informatics 2018, 3203, 1–11.
[17] Huang, Y.; Dang, Z.; Choi, Y.; Andrade, J.; Barilan, A. High-precision smart system on accelerometers and inclinometers for Structural Health Monitoring: Development and applications. In Proceedings of the 2018 12th France-Japan and 10th Europe-Asia Congress on Mechatronics, Tsu, Japan, 10–12 September 2018; pp. 52–57.
[18] Wang, Y.; Ram, S.; Currim, F.; Dantas, E.; Sabóia, L.A. A big data approach for smart transportation management on bus network. In Proceedings of the IEEE 2nd International Smart City Conference: Improving the Citizens Quality of Life, ISC2 2016—Proceedings, Trento, Italy, 12–15 September 2016; pp. 1–6.

[19] Tao, F.; Cheng, J.; Qi, Q. IIHub: An industrial internet-of-things hub toward smart manufacturing based on cyber-physical system. IEEE Trans. Ind. Inform. 2018, 14, 2271–2280.

[20] Trakadas, P.; Simoens, P.; Gkonis, P.; Sarakis, L.; Angelopoulos, A.; Ramallo-González, A.P.; Skarmeta, A.; Trochoutsos, C.; Calvo, D.; Pariente, T.; et al. An artificial intelligence-based collaboration approach in industrial iot manufacturing: Key concepts, architectural extensions and potential applications. Sensors 2020, 20, 5480.

[21] Wan, J.; Yang, J.; Wang, Z.; Hua, Q. Artificial intelligence for cloud-assisted smart factory. IEEE Access 2018, 6, 55419–55430.

[22] Andreão, R.V.; Athayde, M.; Boudy, J.; Aguilar, P.; de Araujo, I.; Andrade, R. Raspcare: A Telemedicine Platform for the Treatment and Monitoring of Patients with Chronic Diseases. In Assistive Technologies in Smart City; IntechOpen: London, UK, 2018. 24. Keane, P.A.; Topol, E.J. With an eye to AI and autonomous diagnosis. NPJ Digit. Med. 2018, 1, 10–12

[23] Trencher, G.; Karvonen, A. Stretching "smart": Advancing health and well-being through the smart city agenda. Local Environ. 2019, 24, 610–627.

[24] de Souza, J.T.; de Francisco, A.C.; Piekarski, C.M.; do Prado, G.F. Data mining and machine learning to promote smart city: A systematic review from 2000 to 2018. Sustainability 2019, 11, 1077.

[25] Rayan, Z.; Alfonse, M.; Salem, A.B.M. Machine Learning Approaches in Smart Health. Procedia Comput. Sci. 2018, 154, 361–368.

[26] Park, J.; Jang, K.; Yang, S.B. Deep neural networks for activity recognition with multi-sensor data in a smart home. In Proceedings of the IEEE World Forum on Internet of Things, WF-IoT 2018—Proceedings, Singapore, 5–8 February 2018; , pp. 155–160.

[27] Wang, A.; Chen, G.; Shang, C.; Zhang, M.; Liu, L. Human activity recognition in a smart home environment with stacked denoising autoencoders. In Proceedings of the International Conference on Web-Age Information Management, Nanchang, China, 3–5 June 2016; pp. 29–40.

[28] Kim, G.Y.; Shin, S.S.; Kim, J.Y.; Kim, H.G. Haptic Conversion Using Detected Sound Event in Home Monitoring System for the Hard-of-Hearing. In Proceedings of the HAVE 2018—IEEE International Symposium on Haptic, Audio-Visual Environments and Games, Proceedings, Dalian, China, 20–21 September 2018; pp. 17–22.

[29] Adege, A.B.; Lin, H.P.; Tarekegn, G.B.; Jeng, S.S. Applying deep neural network (DNN) for robust indoor localization in multi-building environment. Appl. Sci. 2018, 8, 62.

[30] Zimmermann, L.; Weigel, R.; Fischer, G. Fusion of nonintrusive environmental sensors for occupancy detection in smart homes. IEEE Internet Things J. 2018, 5, 2343–2352.
[31] Chowdhry, D.; Paranjape, R.; Laforge, P. Smart home automation system for intrusion detection. In Proceedings of the 2015 IEEE 14th Canadian Workshop on Information Theory, CWIT 2015, St. John's, NL, Canada, 6–9 July 2015; pp. 75–78
[32] Jalal, D. Toward a Smart Real Time Monitoring System for Drinking Water Based on Machine Learning. In Proceedings of the 2019 International Conference on Software, Telecommunications and Computer Networks (SoftCOM), Split, Croatia, 19–21 September 2019; pp. 1–5.
[33] Bin, T.; Alam, M.M.; Absar, N.; Andersson, K.; Shahadat, M. IoT Based Real-time River Water Quality Monitoring System. Procedia Comput. Sci.2019, 155, 161–168
[34] Rosero-Montalvo, P.D.; Caraguay-Procel, J.A.; Jaramillo, E.D.; Michilena-Calderon, J.M.; Umaquinga-Criollo, A.C.; Mediavilla Valverde, M.; Ruiz, M.A.; Beltran, L.A.; Peluffo-Ordónez, D.H. Air quality monitoring intelligent system using machine learning techniques. In Proceedings of the 3rd International Conference on Information Systems and Computer Science, INCISCOS 2018, Quito, Ecuador, 14–16 November 2018; pp. 75–80.
[35] Kök, I.; ¸Sim¸sek, M.U.; Özdemir, S. A deep learning model for air quality prediction in smart city. In Proceedings of the Proceedings—2017 IEEE International Conference on Big Data, Big Data 2017, Boston, MA, USA, 11–14 December 2017; pp. 1983–1990.
[36] Al-Wakeel, A.; Wu, J.; Jenkins, N. K -Means Based Load Estimation of Domestic Smart Meter Measurements. Appl. Energy 2017, 194, 333–342.
[37] Vrablecová, P.; Bou Ezzeddine, A.; Rozinajová, V.; Šárik, S.; Sangaiah, A.K. Smart grid load forecasting using online support vector regression. Comput. Electr. Eng. 2018, 65, 102–117.
[38] Kong, W.; Dong, Z.Y.; Jia, Y.; Hill, D.J.; Xu, Y.; Zhang, Y. Short-Term Residential Load Forecasting Based on LSTM Recurrent Neural Network. IEEE Trans. Smart Grid 2019, 10, 841–851.
[39] Hosein, S.; Hosein, P. Load forecasting using deep neural networks. In Proceedings of the 2017 IEEE Power and Energy Society Innovative Smart Grid Technologies Conference, ISGT 2017, Washington, DC, USA, 23–26 April 2017.

[40] Nguyen, D.; Barella, R.; Wallace, S.A.; Zhao, X.; Liang, X. Smart grid line event classification using supervised learning over PMU data streams. In Proceedings of the 2015 6th International Green and Sustainable Computing Conference, Las Vegas, NV, USA, 14–16 December 2015.

[41] Zheng, Z.; Yang, Y.; Niu, X.; Dai, H.N.; Zhou, Y. Wide and Deep Convolutional Neural Networks for Electricity-Theft Detection to Secure Smart Grids. IEEE Trans. Ind. Infor. 2018, 14, 1606–1615.

[42] Diez, A.; Khoa, N.L.D.; Makki Alamdari, M.; Wang, Y.; Chen, F.; Runcie, P. A clustering approach for structural health monitoring on bridges. J. Civ. Struct. Health Monit. 2016, 6, 429–445.

[43] Guo, J.; Xie, X.; Bie, R.; Sun, L. Structural health monitoring by using a sparse coding-based deep learning algorithm with wireless sensor networks. Pers. Ubiquitous Comput. 2014, 18, 1977–1987.

[44] Dang, H.V.; Tran-Ngoc, H.; Nguyen, T.V.; Bui-Tien, T.; De Roeck, G.; Nguyen, H.X. Data-driven structural health monitoring using feature fusion and hybrid deep learning. IEEE Trans. Autom. Sci. Eng. 2020.

[45] Vitola, J.; Pozo, F.; Tibaduiza, D.A.; Anaya, M. A sensor data fusion system based on k-nearest neighbor pattern classification for structural health monitoring applications. Sensors 2017, 17, 417.

[46] Yu, J.; Kim, M.; Bang, H.C.; Bae, S.H.; Kim, S.J. IoT as a applications: cloud-based building management systems for the internet of things. Multimed. Tools Appl. 2016, 75, 14583–14596.

[47] Singaravel, S.; Geyer, P.; Suykens, J. Deep neural network architectures for component-based machine learning model in building energy predictions. In Proceedings of the Digital Proceedings of the 24th EG-ICE International Workshop on Intelligent Computing in Engineering, Nottingham, UK, 10–12 July 2017.

[48] Hu, W.; Wen, Y.; Guan, K.; Jin, G.; Tseng, K.J. itcm: Toward learning-based thermal comfort modeling via pervasive sensing for smart buildings. IEEE Internet Things J. 2018, 5, 4164–4177.

[49] Sevillano, X.; Màrmol, E.; Fernandez-Arguedas, V. Towards smart traffic management systems: Vacant on-street parking spot detection based on video analytics. In Proceedings of the FUSION 2014—17th International Conference on Information Fusion, Salamanca, Spain, 7–10 July 2014.

[50] Stolfi, D.H.; Alba, E.; Yao, X. Predicting car park occupancy rates in smart city. In Proceedings of the International Conference on Smart City, Malaga, Spain, 14–16 June 2017 ; pp. 107–117.

[51] Ali, G.; Ali, T.; Irfan, M.; Draz, U.; Sohail, M.; Glowacz, A.; Sulowicz, M.; Mielnik, R.; Faheem, Z.B.; Martis, C. IoT Based Smart Parking System Using Deep Long Short Memory Network. Electronics 2020, 9, 1696.

[52] Ebuchi, T.; Yamamoto, H. Vehicle/pedestrian localization system using multiple radio beacons and machine learning for smart parking. In Proceedings of the 2019 International Conference on Artificial Intelligence in Information and Communication (ICAIIC), Okinawa, Japan, 11–13 February 2019; pp. 086–091.

[53] Sun, F.; Pan, Y.; White, J.; Dubey, A. Real-Time and Predictive Analytics for Smart Public Transportation Decision Support System. In Proceedings of the 2016 IEEE International Conference on Smart Computing, SMARTCOMP 2016, St Louis, MO, USA, 18–20 May 2016; pp. 1–8.

[54] Xiao, Y.; Yin, Y. Hybrid LSTM neural network for short-term traffic flow prediction. Information 2019, 10, 105.

[55] Wei, W.; Wu, H.; Ma, H. An autoencoder and LSTM-based traffic flow prediction method. Sensors 2019, 19, 2946.

[56] Castro, D.; Coral, W.; Rodriguez, C.; Cabra, J.; Colorado, J. Wearable-based human activity recognition using an iot approach. J. Sens. Actuator Netw. 2017, 6, 28.

[57] Santos, G.L.; Endo, P.T.; Monteiro, K.H.d.C.; Rocha, E.d.S.; Silva, I.; Lynn, T. Accelerometer-based human fall detection using convolutional neural networks. Sensors 2019, 19, 1644.

[58] Tran, D.N.; Phan, D.D. Human Activities Recognition in Android Smartphone Using Support Vector Machine. In Proceedings of the International Conference on Intelligent Systems, Modelling and Simulation, ISMS, Bangkok, Thailand, 25–27 January 2016; pp. 64–68.

[59] Alamri, A. Monitoring System for Patients Using Multimedia for Smart Healthcare. IEEE Access 2018, 6, 23271–23276.

[60] Awais, M.; Raza, M.; Singh, N.; Bashir, K.; Manzoor, U.; ul Islam, S.; Rodrigues, J.J. LSTM based Emotion Detection using Physiological Signals: IoT framework for Healthcare and Distance Learning in COVID-19. IEEE Internet Things J. 2020.

[61] Alhussein, M.; Muhammad, G.; Hossain, M.S.; Amin, S.U. Cognitive IoT-Cloud Integration for Smart Healthcare: Case Study for Epileptic Seizure Detection and Monitoring. Mob. Netw. Appl. 2018, 23, 1624–1635.

[62] Bhide, V.H.; Wagh, S. I-learning IoT: An intelligent self learning system for home automation using IoT. In Proceedings of the 2015 International

Conference on Communication and Signal Processing, ICCSP 2015, Melmaruvathur, India, 2–4 April 2015; pp. 1763–1767.

[63] Mohapatra, S.; Patra, P.K.; Mohanty, S.; Pati, B. Smart Health Care System using Data Mining. In Proceedings of the 2018 International Conference on Information Technology, ICIT 2018, Bhubaneswar, India, 19–21 December 2018; pp. 44–49.

[64] Kaur, P.; Kumar, R.; Kumar, M. A healthcare monitoring system using random forest and internet of things (IoT). Multimed. Tools Appl. 2019, 78, 19905–19916.

[65] Tuli, S.; Basumatary, N.; Gill, S.S.; Kahani, M.; Arya, R.C.; Wander, G.S.; Buyya, R. HealthFog: An ensemble deep learning based Smart Healthcare System for Automatic Diagnosis of Heart Diseases in integrated IoT and fog computing environments. Future Gener. Comput. Syst. 2020, 104, 187–200.

[66] Ani, R.; Krishna, S.; Anju, N.; Sona, A.M.; Deepa, O.S. IoT based patient monitoring and diagnostic prediction tool using ensemble classifier. In Proceedings of the 2017 International Conference on Advances in Computing, Communications and Informatics, ICACCI 2017, Udupi (Near Mangalore), India, 13–16 September 2017; pp. 1588–1593.

[67] Alhussein, M. Monitoring Parkinson's Disease in Smart City. IEEE Access 2017, 5, 19835–19841.

[68] Devarajan, M.; Ravi, L. Intelligent cyber-physical system for an efficient detection of Parkinson disease using fog computing. Multimed. Tools Appl. 2019, 78, 32695–32719.

[69] Sayeed, M.A.; Mohanty, S.P.; Kougianos, E.; Yanambaka, V.P.; Zaveri, H. A robust and fast seizure detector for IoT edge. In Proceedings of the 2018 IEEE 4th International Symposium on Smart Electronic Systems, iSES 2018, Hyderabad, India, 17–19 December2018; pp. 156–160.

[70] Varghese, R.; Sharma, S. Affordable smart farming using IoT and machine learning. In Proceedings of the 2018 Second International Conference on Intelligent Computing and Control Systems (ICICCS), Madurai, India. 14–15 June 2018; pp. 645–650.

[71] Pratyush Reddy, K.S.; Roopa, Y.M.; Kovvada Rajeev, L.N.; Nandan, N.S. IoT based Smart Agriculture using Machine Learning. In Proceedings of the 2nd International Conference on Inventive Research in Computing Applications, ICIRCA 2020, Coimbatore, India, 15–17 July 2020; pp. 130–134.

[72] AlZu'bi, S.; Hawashin, B.; Mujahed, M.; Jararweh, Y.; Gupta, B.B. An efficient employment of internet of multimedia things in smart and future agriculture. Multimed. Tools Appl. 2019, 78, 29581–29605.

[73] Goap, A.; Sharma, D.; Shukla, A.K.; Rama Krishna, C. An IoT based smart irrigation management system using Machine learning and open source technologies. Comput. Electron. Agric. 2018, 155, 41–49.

[74] Rodríguez, S.; Gualotuña, T.; Grilo, C. A System for the Monitoring and Predicting of Data in Precision A Agriculture System for the and Predicting of Wireless Data in Precision in a Monitoring Rose Greenhouse Based on Sensor Agriculture in a Rose Greenhouse Based on Wireless Sensor Networks Ne. Procedia Comput. Sci. 2017, 121, 306–313.

[75] Kitpo, N.; Kugai, Y.; Inoue, M.; Yokemura, T.; Satomura, S. Internet of Things for Greenhouse Monitoring System Using Deep Learning and Bot Notification Services. In Proceedings of the 2019 IEEE International Conference on Consumer Electronics, ICCE 2019, Las Vegas, NV, USA, 11–13 January 2019.

[76] Saha, A.K.; Saha, J.; Ray, R.; Sircar, S.; Dutta, S.; Chattopadhyay, S.P.; Saha, H.N. IOT-based drone for improvement of crop quality in agricultural field. In Proceedings of the 2018 IEEE 8th Annual Computing and Communication Workshop and Conference (CCWC), Las Vegas, NV, USA, 8–10 January 2018; pp. 612–615.

[77] Araby, A.A.; Abd Elhameed, M.M.; Magdy, N.M.; Abdelaal, N.; Abd Allah, Y.T.; Darweesh, M.S.; Fahim, M.A.; Mostafa, H. Smart iot monitoring system for agriculture with predictive analysis. In Proceedings of the 2019 8th International Conference on Modern Circuits and Systems Technologies (MOCAST), Thessaloniki, Greece, 13–15 May 2019; pp. 1–4.

[78] Nandhini, S.A.; Radha, R.H.S. Web Enabled Plant Disease Detection System for Agricultural Applications Using WMSN.Wirel.Pers. Commun. 2018, 102, 725–740.

[79] Ale, L.; Sheta, A.; Li, L.; Wang, Y.; Zhang, N. Deep learning based plant disease detection for smart agriculture. In Proceedings of the 2019 IEEE Globecom Workshops, GC Wkshps 2019—Proceedings, Waikoloa, HI, USA, 9–13 December 2019; pp. 1–6.

[80] Jin, X.B.; Yang, N.X.; Wang, X.Y.; Bai, Y.T.; Su, T.L.; Kong, J.L. Hybrid deep learning predictor for smart agriculture sensing based on empirical mode decomposition and gated recurrent unit group model. Sensors 2020, 20, 1334.

[81] Diedrichs, A.L.; Bromberg, F.; Dujovne, D.; Brun-Laguna, K.; Watteyne, T. Prediction of frost events using machine learning and IoT sensing devices. IEEE Internet Things J. 2018, 5, 4589–4597.

[82] Balducci, F.; Impedovo, D.; Pirlo, G. Machine learning applications on agricultural datasets for smart farm enhancement. Machines 2018, 6,38.

[83] Aliev, K.; Jawaid, M.M.; Narejo, S.; Pasero, E.; Pulatov, A. Internet of plants application for smart agriculture. Int. J. Adv. Comput. Sci. Appl. 2018, 9, 421–429.

[84] Chalee Vorakulpipat, Ryan K. L. Ko, Qi Li, and Ahmed Meddahi, Security and Privacy in Smart City, Hindawi Security and Communication Networks Volume 2021, Article ID 9830547, 2 pages https://doi.org/10.1155/2021/9830547, Published 15 August 2021.

[85] Saini, A.; Malik, A. Routing in internet of things: A survey. Commun. Comput. Syst. 2017, I, 855–860.

[86] Risteska Stojkoska, B.L.; Trivodaliev, K.V. A review of Internet of Things for smart home: Challenges and solutions. J. Clean. Prod. 2017, 140, 1454–1464.

[87] Mehmood, Y.; Ahmad, F.; Yaqoob, I.; Adnane, A.; Imran, M.; Guizani, S. Internet-of-Things-Based Smart City: Recent Advances and Challenges. IEEE Commun. Mag. 2017, 55, 16–24.

[88] Rong, W.; Xiong, Z.; Cooper, D.; Li, C.; Sheng, H. Smart city architecture: A technology guide for implementation and design challenges. China Commun. 2014, 11, 56–69.

[89] Bar-Magen Numhauser, J. Fog Computing Introduction to a New Cloud Evolution; University of Alcalá: Alcalá de Henares, Spain, 2012.

[90] Aazam, M.; Zeadally, S.; Harras, K.A. Fog Computing Architecture, Evaluation, and Future Research Directions. IEEE Commun. Mag. 2018, 56, 46–52.

[91] El-Sayed, H.; Sankar, S.; Prasad, M.; Puthal, D.; Gupta, A.; Mohanty, M.; Lin, C.T. Edge of Things: The Big Picture on the Integration of Edge, IoT and the Cloud in a Distributed Computing Environment. IEEE Access 2017, 6, 1706–1717.

Index

About the Editors

M.A. Jabbar is a Professor and Head of the Department AI&ML, Vardhaman College of Engineering, Hyderabad and Telangana, India. He obtained his Doctor of Philosophy (PhD) from JNTUH, Hyderabad and Telangana, India. He has been teaching for more than 20 years. His research interests include artificial intelligence, big data analytics, bio-informatics, cyber security, machine learning, attack graphs, and intrusion detection systems. He has published more than 57 papers in various journals and conferences. He has served as a technical committee member for more than 70 international conferences, and has edited 5 books with various leading publishers. He is a senior Member of IEEE and Senior member of ACM, Governing body member, Internet Society India Hyderabad Chapter. Presently he is acting as a Chair, IEEE CS chapter, Hyderabad Section.

Sanju Tiwari is a Senior Researcher at Universidad Autonoma de Tamaulipas (70-year-old University), Mexico. She is a DAAD Post-Doc-Net AI Fellow for 2021. She is an Adjunct Professor at Vardhman Engineering College, Hyderabad in the Computer Science department. She has also been appointed as PhD co-supervisor at Rai University, Gujarat, India. She is a senior member of IEEE as a SMIEEE. She has worked as a Post-Doctoral Researcher in the Ontology Engineering Group, Universidad Polytecnica De Madrid, Spain. Prior to this, she worked as a Research Associate for a sponsored research project "Intelligent Real-time Situation Awareness and Decision Support System for Indian Defence" funded by DRDO, New Delhi in the Department of Computer Applications, National Institute of Technology, Kurukshetra. In this project, she has developed and evaluated a Decision Support System for Indian Defence. Her current research interests include ontology engineering, knowledge graphs, linked data generation and publication, semantic web, reasoning with SPARQL, and machine intelligence.

Fernando Ortiz-Rodriguez is a full professor at Tamaulipas Autonomous University, Member of National Research Council level C, and director of the Social Research Centre at UAT, Mexico. He is a member of the Information

Technology research group and part of the Knowledge Graph and Semantic Web Community. He has published, journal articles, book chapters, and has acted as a book editor for reputed editorials. He has also presented many conference articles in Europe, the United States, and Mexico and is the main chair and organizer of the KGSWC multiseries conference.

For Product Safety Concerns and Information please contact our EU
representative GPSR@taylorandfrancis.com
Taylor & Francis Verlag GmbH, Kaufingerstraße 24, 80331 München, Germany

www.ingramcontent.com/pod-product-compliance
Ingram Content Group UK Ltd.
Pitfield, Milton Keynes, MK11 3LW, UK
UKHW021110180425
457613UK00001B/14
9788770227490